Charles Whymper

The Gamekeeper at Home

Sketches of Natural History and Rural Life

Charles Whymper

The Gamekeeper at Home
Sketches of Natural History and Rural Life

ISBN/EAN: 9783744716642

Printed in Europe, USA, Canada, Australia, Japan

Cover: Foto ©berggeist007 / pixelio.de

More available books at **www.hansebooks.com**

THE

GAMEKEEPER AT HOME

SKETCHES OF NATURAL HISTORY
AND RURAL LIFE

When shaws beene sheene, and shradds full fayre,
    And leaves both large and longe,
It is merrye walking in the fayre forrest,
    To hear the small birdes songe.

Ballad of Guy of Gisborne.

WITH ILLUSTRATIONS BY CHARLES WHYMPER

𝕷𝖔𝖓𝖉𝖔𝖓

SMITH, ELDER, & CO., 15 WATERLOO PLACE

1880

PREFACE.

THOSE who delight in roaming about amongst the fields and lanes, or have spent any time in a country house, can hardly have failed to notice the custodian of the woods and covers, or to observe that he is often something of a 'character.' The Gamekeeper forms, indeed, so prominent a figure in rural life as almost to demand some biographical record of his work and ways. From the man to the territories over which he bears sway—the meadows, woods, and streams—and to his subjects, their furred and feathered inhabitants, is a natural transition. The enemies against whom he wages incessant warfare—vermin, poachers, and trespassers—must, of course, be included in such a survey.

Although, for ease and convenience of illustra-tion, the character of a particular Keeper has been used as a nucleus about which to arrange materials that would otherwise have lacked a connecting link, the facts here collected are really entirely derived from original observation.

R. J.

CONTENTS.

LIST OF ILLUSTRATIONS.

THE KEEPER'S COTTAGE.

CHAPTER I.

The Man himself—his House, and Tools.

THE keeper's cottage stands in a sheltered 'coombe,' or narrow hollow of the woodlands, overshadowed by a mighty Spanish chestnut, bare now of leaves, but in summer a noble tree. The ash wood covers the slope at the rear ; on one side is a garden, and on the other a long strip of meadow with elms. In front, and somewhat lower, a streamlet winds, fringing the sward, and across it

B

the fir plantations begin, their dark sombre foliage hanging over the water. A dead willow trunk thrown from bank to bank forms a rude bridge ; the tree, not even squared, gives little surface for the foot, and in frosty weather a slip is easy. From this primitive contrivance a path, out of which others fork, leads into the intricacies of the covers, and from the garden a wicket-gate opens on the ash wood. The elms in the meadow are full of rooks' nests, and in the spring the coombe will resound with their cawing ; these black bandits, who do not touch it at other times, will then ravage the garden to feed their hungry young, despite ingenious scarecrows. A row of kennels, tenanted by a dozen dogs, extends behind the cottage : lean retrievers yet unbroken, yelping spaniels, pointers, and perhaps a few greyhounds or fancy breeds, if ' young master ' has a taste that way.

Beside the kennels is a shed ornamented with rows upon rows of dead and dried vermin, furred and feathered, impaled for their misdeeds ; and over the door a couple of horseshoes nailed for luck—a superstition yet lingering in the by-ways of the woods and hills. Within are the ferret hutches, warm and dry ; for the ferret is a shivery creature, and likes nothing so well as to nozzle down in a coat-pocket with a little hay. Here are spades and bill-hooks, twine and rabbit nets, traps, and other odds and ends scattered about with the wires and poacher's implements impounded from time to time.

In a dark corner there lies a singular-looking piece of mechanism, a relic of the olden times, which, when dragged into the light, turns out to be a man-trap. These terrible engines have long since been disused—being illegal, like

THE ROW OF KENNELS.

spring-guns—and the rust has gathered thickly on the metal. But, old though it be, it still acts perfectly, and can be 'set' as well now as when in bygone days poachers and thieves used to prod the ground and the long grass, before they stepped among it, with a stick, for fear of mutilation.

The trap is almost precisely similar to the common

rat-trap or gin still employed to destroy vermin, but greatly exaggerated in size, so that if stood on end it reaches to the waist, or above.   The jaws of this iron wolf are horrible to contemplate—rows of serrated projections, which fit into each other when closed, alternating with spikes a couple of inches long, like tusks.   To set the trap you have to stand on the spring—the weight of a man is about sufficient to press it down ; and, to avoid danger to the person preparing this little surprise, a band of iron can be pushed forward to hold the spring while the catch is put into position, and the machine itself is hidden among the bushes or covered with dead leaves.   Now touch the pan with a stout walking-stick—the jaws cut it in two in the twinkling of an eye.   They seem to snap together with a vicious energy, powerful enough to break the bone of the leg ; and assuredly no man ever got free whose foot was once caught by these terrible teeth.

The keeper will tell you that it used to be set up in the corner of the gardens and orchard belonging to the great house, and which, in the pre-policemen days, were almost nightly robbed.   He thinks there were quite as many such traps set in the gardens just outside the towns as ever there were in the woods and preserves of the country proper.   He recollects but one old man (a mole-catcher) who actually had experienced in his youth the sensation of being caught ; he went lame on one foot, the sinews having been cut or divided.   The trap could be chained to its place if desired ; but, as a matter of fact, a chain was unnecessary, for no man could possibly drag this torturing clog along.

Another outhouse attached to the cottage contains a copper for preparing the food for both quadrupeds and birds. Some poultry run about the mead, and perhaps with them are feeding the fancy foreign ducks which in summer swim in the lake before the hall.

The cottage is thatched and oddly gabled—built before 'improvements' came into fashion—yet cosy ; with walls three feet thick, which keep out the cold of winter and the heat of summer. This is not solid masonry ; there are two shells, as it were, filled up between with rubble and mortar rammed down hard.

Inside the door the floor of brick is a step below the level of the ground. Sometimes a peculiar but not altogether unpleasant odour fills the low-pitched sitting-room —it is emitted by the roots burning upon the fire, hissing as the sap exudes and boils in the fierce heat. When the annual fall of timber takes place the butts of the trees are often left in the earth, to be afterwards grubbed and split for firewood, which goes to the great house or is sold. There still remain the roots, which are cut into useful lengths and divided among the upper employés. From elm and oak and ash, and the crude turpentine of the fir, this aromatic odour, the scent of the earth in which they grew, is exhaled as they burn.

THE KEEPER'S KITCHEN.

The ceiling is low and crossed by one huge square beam of oak, darkened by smoke and age. The keeper's double-barrelled gun is suspended from this beam : there are several other guns in the house, but this, the favourite, alone hangs where it did before he had children—so strong is habit ; the rest are yet more out of danger. It has been a noble weapon, though now showing signs of age— the interior of the breech worn larger than the rest of the barrel from constant use ; so much so that, before it was converted to a breech-loader, the wad, when the ramrod pushed it down, would slip the last six inches, so loosely fitting as to barely answer its purpose of retaining the shot ; so that when cleaned out, before the smoke fouled it again, he had to load with paper. This in a measure anticipated the 'choke-bore,' and his gun was always famous for its killing power. The varnish is worn from the stock by incessant friction against his coat, showing the real grain of the walnut-wood, and the trigger-guard with the polish of the sleeve shines like silver. It has been his companion for so many years that it is not strange he should feel an affection for it ; no other ever fitted the shoulder so well, or came with such delicate precision to the 'present' position. So accustomed is he to its balance and 'hang' in the hand that he never thinks of aiming ; he simply looks at the object, still or moving, throws the gun up from the hollow of his arm, and instantly pulls the trigger, staying not a second to glance along the barrel. It has become almost a portion of his body, answering like a limb to the volition of will without the intervention of reflection. The hammers are chased and elegantly shaped—perfectly matching : when once the screw came loose, and the jar of a shot jerked one off among the dead leaves apparently beyond hope of recovery, he never rested night or day till by continuous search and sifting the artistic piece of metal was found. Nothing destroys the symmetry of a gun so much as hammers which are not pairs ; and well he knew that he should

never get a smith to replace that delicate piece of work-
manship, for this gun came originally from the hands of
a famous maker, who got fifty, or perhaps even seventy
guineas for it years ago.   It did not shoot to please the
purchaser—guns of the very best character sometimes take
use to get into thorough order—and was thrown aside,
and so the gun became the keeper's.

These fine old guns often have a romance clinging to
them, and sometimes the history is a sad one.   Upstairs
he still keeps the old copper powder-flask curiously chased
and engraved, yet strong enough to bear the weight of the
bearer if by chance he sat down upon it while in his
pocket, together with the shot-belt and punch for cutting
out the wads from card-board or an old felt hat.   These
the modern system of loading at the breech has cast aside.
Here, also, is the apparatus for filling empty cartridge-
cases—a work which in the season occupies him many
hours.

Being an artist in his way, he takes a pride in the
shine and polish of his master's guns, which are not always
here, but come down at intervals to be cleaned and
attended to.   And woe be to the first kid gloves that
touch them afterwards ; for a gun, like a sardine, should
be kept in fine oil, not thickly encrusting it, but, as it
were, rubbed into and oozing from the pores of the metal
and wood.   Paraffin is an abomination in his eyes (for
preserving from rust), and no modern patent oil, he thinks,

can compare with a drop of gin for the locks—the spirit never congeals in cold weather, and the hammer comes up with a clear, sharp snick. He has two or three small screwdrivers and gunsmith's implements to take the locks to pieces ; for gentlemen are sometimes careless and throw their guns down on the wet grass, and if a single drop of water should by chance penetrate under the plate it will play mischief with the works if the first speck of rust be not forthwith removed.

His dog-whistle hangs at his buttonhole. His pocket-knife is a basket of tools in itself, most probably a present from some youthful sportsman who was placed under his care to learn how to handle a gun. The corkscrew it contains has seen much service at luncheon-time, when under a sturdy oak, or in a sheltered nook of the lane, where the hawthorn hedge and the fern broke the force of the wind, a merry shooting-party sat down to a well-packed hamper and wanted some one to draw the corks. Not but what the back of the larger blade has not artistic-ally tapped off the neck of many a bottle, hitting it gently upwards against the rim. Nor must his keys be forgotten. The paths through the preserves, where they debouch on a public lane or road, are closed with high-sparred wicket gates, well pitched to stand the weather, and carefully locked, and of course he has a key. His watch, made on purpose for those who walk by night, tells him the time in the densest darkness of the woods. On pressing a

spring and holding it near the ear, it strikes the hour last
passed, then the quarters which have since elapsed ; so
that even when he cannot see an inch before his face he
knows the time within fifteen minutes at the outside, which
is near enough for practical purposes.

In personal appearance he would be a tall man were
it not that he has contracted a slight stoop in the passage
of the years, not from weakness or decay of nature, but
because men who walk much lean forward somewhat,
which has a tendency to round the shoulders.   The
weight of the gun, and often of a heavy game-bag drag-
ging downwards, has increased this defect of his figure,
and, as is usual after a certain age, even with those who
lead a temperate life, he begins to show signs of corpulency.
But these shortcomings only slightly detract from the
manliness of his appearance, and in youth it is easy to see
that he must have been an athlete.   There is still plenty
of power in the long sinewy arms, brown hands, and bull-
neck, and intense vital energy in the bright blue eye.
He is an ash-tree man, as a certain famous writer would
say ; hard, tough, unconquerable by wind or weather,
fearless of his fellows, yielding but by slow and imper-
ceptible degrees to the work of time.   His neck has
become the colour of mahogany ; sun and tempest have
left their indelible marks upon his face ; and he speaks
from the depths of his broad chest, as men do who
talk much in the open air, shouting across the fields

and through the copses. There is a solidity in his very footstep, and he stands like an oak. He meets your eye

THE KEEPER.

full and unshirkingly, yet without insolence ; not as the labourers do, who either stare with sullen ill-will or look on the earth. In brief, freedom and constant contact with

nature have made him every inch a man ; and here in this
nineteenth century of civilised effeminacy may be seen
some relic of what men were in the old feudal days when
they dwelt practically in the woods.   The shoulder of his
coat is worn a little where the gun rubs, and so is his
sleeve ; otherwise he is fairly well dressed.

Perfectly civil to every one, and with a willing
manner towards his master and his master's guests, he
has a wonderful knack of getting his own way.   Whatever
the great house may propose in the shooting line, the
keeper is pretty certain to dispose of in the end as he
pleases ; for he has a voluble 'silver' tongue, and is full
of objections, reasons, excuses, suggestions, all delivered
with a deprecatory air of superior knowledge which he
hardly likes to intrude upon his betters, much as he would
regret to see them go wrong.   So he really takes the lead,
and in nine cases in ten the result proves he is right, as
minute local knowledge naturally must be when intelli-
gently applied.

Not only in such matters as the best course for the
shooting-party to follow, or in advice bearing upon the
preserves, but in concerns of a wider scope, his influence
is felt.   A keen, shrewd judge of horseflesh—(how is it
that if a man understands one animal he seems to
instinctively see through all ?)—his master in a careless
way often asks his opinion before concluding a bargain.
Of course the question is not put direct, but ' By-the-bye,

when the hounds were here you saw so-and-so's mare ;
what do you think of her ? ' The keeper blurts out his
answer, not always flattering or very delicately expressed ;
and his view is not forgotten. For when a trusted servant
like this accompanies his master often in solitary rambles
for hours together, dignity must unbend now and then,
however great the social difference between them ; and
thus a man of strong individuality and a really valuable
gift of observation insensibly guides his master.

Passing across the turnips, the landlord, who perhaps
never sees his farms save when thus crossing them with
a gun, remarks that they look clean and free from weeds ;
whereupon the keeper, walking respectfully a little in the
rear, replies that so-and-so, the tenant, is a capital farmer,
a preserver of foxes and game, but has suffered from the
floods—a reply that leads to inquiries, and perhaps a
welcome reduction of rent. On the other hand, the
owner's attention is thus often called to abuses. In this
way an evilly-disposed keeper may, it is true, do great
wrongs, having access to the owner and, in familiar phrase,
' his ear.' I am at present delineating the upright keeper,
such as are in existence still, notwithstanding the abuse
lavished upon them as a class—often, it is to be feared,
too well deserved. It is not difficult to see how in this
way a man whose position is lowly may in an indirect
way exercise a powerful influence upon a large estate.

He is very ' great ' on dogs (and, indeed, on all other

animals) ; his opinion is listened to and taken by everybody round about who has a dog, and sometimes he has
three or four under treatment for divers ills.   By this
knowledge many 'tips' are gained, and occasionally he
makes a good thing by selling a pup at a high price.   He
may even be seen, with his velveteen jacket carefully
brushed, his ground-ash stick under his arm, and hat in
hand, treading daintily for fear of soiling the carpet with
his shoes, in the ante-room, gravely prescribing for the
ailing pug in which the ladies are interested.

At the farmhouses he is invited to sit down and take
a glass, being welcome for his gossip of the great house,
and because, having in the course of years been thrown
into the society of all classes, he has gradually acquired a
certain tact and power of accommodating himself to his
listener.   For the keeper, when he fulfils his duty in a
quiet way, as a man of experience does, is by no means
an unpopular character.   It is the too officious man who
creates a feeling among the tenants against himself and
the whole question of game.   But the quiet experienced
hand, with a shrewd knowledge of men as well as the
technicalities of his profession, grows to be liked by the
tenantry, and becomes a local authority on animal life.

Proud, and not without reason, of his vigour and
strength, he will tell you that though between fifty and
sixty he can still step briskly through a heavy field-day,
despite the weight of reserve ammunition he carries.   He

can keep on his feet without fatigue from morn till eve, and goes his rounds without abating one inch of the distance. In one thing alone he feels his years—*i.e.* in pace ; and when 'young master,' who is a disciple of the modern athletic school, comes out, it is about as much as ever he can do to keep up with him over the stubble. Never once for the last thirty years has he tossed on a bed of sickness ; never once has he failed to rise from his slumber refreshed and ready for his labour. His secret is—but let him tell it in his own words :

'It's indoors, sir, as kills half the people ; being indoors three parts of the day, and next to that taking too much drink and vittals. Eating's as bad as drinking ; and there ain't nothing like fresh air and the smell of the woods. You should come out here in the spring, when the oak timber is throwed (because, you see, the sap be rising, and the bark strips then), and just sit down on a stick fresh peeled—I means a trunk, you know—and sniff up the scent of that there oak bark. It goes right down your throat, and preserves your lungs as the tan do leather. And I've heard say as folk who work in the tan-yards never have no illness. There's always a smell from trees, dead or living—I could tell what wood a log was in the dark by my nose ; and the air is better where the woods be. The ladies up in the great house sometimes goes out into the fir plantations—the turpentine scents strong, you see—and they say it's good for the chest ; but,

bless you, you must live in it. People go abroad, I'm
told, to live in the pine forests to cure 'em : I say these
here oaks have got every bit as much good in that way.
I never eat but two meals a day—breakfast and supper :
what you would call dinner—and maybe in the middle of
the day a hunch of dry bread and an apple. I take a
deal for breakfast, and I'm rather lear [hungry] at supper ;
but you may lay your oath that's why I'm what I am in
the way of health. People stuffs theirselves, and by
consequence it breaks out, you see. It's the same with
cattle ; they're overfed, tied up in stalls and stuffed, and
never no exercise, and mostly oily food too. It stands to
reason they must get bad ; and that's the real cause of
these here rinderpests and pleuro-pneumonia and what-
nots. At least that's my notion. I'm in the woods all
day, and never comes home till supper—'cept, of course,
in breeding-time, to fetch the meal and stuff for the
birds—so I gets the fresh air, you see ; and the fresh air
is the life, sir. There's the smell of the earth, too—
'specially just as the plough turns it up—which is a fine
thing ; and the hedges and the grass are as sweet as
sugar after a shower. Anything with a green leaf is the
thing, depend upon it, if you want to live healthy. I
never signed no pledge ; and if a man asks me to take
a glass of ale, I never says him no. But I ain't got no
barrel at home ; and all the time I've been in this here
place I've never been to a public. Gentlemen give me

tips—of course they does ; and much obliged I be ; but I takes it to my missus. Many's the time they've axed me to have a glass of champagne or brandy when we've had lunch under the hedge ; but I says no, and would like a glass of beer best, which I gets, of course. No ; when I drinks, I drinks ale : but most in general I drinks no strong liquor. Great coat !—cold weather ! I never put no great coat on this thirty year. These here woods be as good as a topcoat in cold weather. Come off the open field with the east wind cutting into you, and get inside they firs and you'll feel warm in a minute. If you goes into the ash wood you must go in farther, because the wind comes more between the poles.' Fresh air, exercise, frugal food and drink, the odour of the earth and the trees—these have given him, as he nears his sixtieth year, the strength and vitality of early manhood.

He has his faults : notably, a hastiness of temper towards his undermen, and towards labourers and wood-cutters who transgress his rules. He is apt to use his ground-ash stick rather freely without thought of con-sequences, and has got into trouble more than once in that way. When he takes a dislike or suspicion of a man, nothing will remove it ; he is stubbornly inimical and unforgiving, totally incapable of comprehending the idea of loving an enemy. He hates cordially in the true pagan fashion of old. He is full of prejudices, and has some ideas which almost amount to superstitions ; and, though

C

he fears nothing, has a vague feeling that sometimes there is 'summat' inexplicable in the dark and desolate places. Such is this modern man of the woods.

The impressions of youth are always strongest with us, and so it is that recollecting the scenes in which he passed his earlier days he looks with some contempt upon the style of agriculture followed in the locality; for he was born in the north, where the farms are sometimes of a great area, though perhaps not so rich in soil, and he cannot forgive the tenants here because they have not got herds of three or four hundred horned cattle. Before he settled down in the south he had many changes of situation, and was thus brought in contact with a wonderful number of gentlemen, titled or otherwise distinguished, whose peculiarities of speech or appearance he loves to dwell upon. If the valet sees the hero or the statesman too closely, so sometimes does the gamekeeper. A great man must have moments when it is a relief to fling off the constant posturing necessary before the world; and there is freshness in the gamekeeper's unstudied conversation. The keeper thinks that nothing reveals a gentleman's character so much as his 'tips.'

'Gentlemen is very curious in tips,' he says, 'and there ain't nothing so difficult as to know what's coming. Most in general them as be the biggest guns, and what you would think would come out handsome, chucks you a crown and no more; and them as you knows ain't much

go in the way of money slips a sovereign into your fist.
There's a deal in the way of giving it too, as perhaps you
wouldn't think. Some gents does it as much as to say
they're much obliged to you for kindly taking it. Some
does it as if they were chucking a bone to a dog. One
place where I was, the governor were the haughtiest man
as ever you see. When the shooting was done—after a
great party, you never knowed whether he were pleased
or not—he never took no more notice of you than if you
were a tree. But I found him out arter a time or two.
You had to walk close behind him, as if you were a spaniel;
and by-and-by he would slip his hand round behind his
back—without a word, mind—and you had to take what
was in it, and never touch your hat or so much as " Thank
you, sir." It were always a five-pound note if the shooting had been good ; but it never seemed to come so sweet
as if he'd done it to your face.'

The keeper gets a goodly number of tips in the course
of the year, from visitors at the great house, from natural-
ists who come now and then, from the sportsmen, and
regularly from the masters of three packs of hounds ; not
to mention odd moneys at intervals in various ways, as
when he goes round to deliver presents of game to the
chief tenants on the estate or to the owner's private friends.
Gentlemen who take an interest in such things come out
every spring to see the young broods of pheasants—which,
indeed, are a pretty sight—and they always leave some-

thing behind them.   In the summer a few picnic parties
come from the town or the country round about, having
permission to enter the grounds.   In the winter half a

A BROOD OF YOUNG PHEASANTS.

dozen young gentlemen have a turn at the ferreting ; a
great burrow is chosen, three or four ferrets put in at once
without any nets, so that the rabbits may bolt freely, and
then the shooting is like volleys of musketry fire.   For
sport like this the young gentlemen tip freely.   After the

rook-shooting party in the spring from the great house, with their rook-rifles and sometimes crossbows, have had the pick of the young birds, some few of the tenants are admitted to shoot the remainder—a task that spreads perhaps over two or even three days, and there is a good deal of liquor and silver going about. Then gentlemen come to fish in the mere, having got the necessary permission, and they want bait and some attention, which the keeper's lad, being an adept himself, can render better than any one else; and so he too gets his share. Besides which, being swift of foot, and with a shrewd idea which way the fox will run when the hunt is up, he is to the fore when a lady or some timid gentleman wants a gate opened—a service not performed in vain. For breaking-in dogs also the keeper is often paid well; and, in short, he is one of those fortunate individuals whom all the world tips.

CHAPTER II.

His Family and Caste.

THE interior of the cottage is exquisitely clean ; it has that bright pleasant appearance which is only possible when the housewife feels a pride in her duties, and goes about them with a cheerful heart. Not a speck of dust can be seen upon the furniture, amongst which is a large old-fashioned sofa : the window panes are clear and transparent—a certain sign of loving care expended on the place, as on the other hand dirty windows are an indication of neglect, so much so that the character of the cottager may almost be guessed from a glance at her glass. The keeper's wife is a buxom vivacious dame, whose manners, from occasional contact with the upper ranks—the ladies from the great house sometimes look in for a few minutes to chat with so old a servant of the family—are above what are usually found in her station. She receives her callers—and they are many—with a quiet, respectful dignity : desirous of pleasing, yet quite at her ease.

Across the back of the sofa there lies a rug of some

beautiful fur which catches the eye, but which at first the visitor cannot identify. Its stripes are familiar, and not unlike the tiger's, but the colour is not that of the forest tyrant. She explains that this rug comes within her special sphere. It is a carriage-rug of cat-skin ; the skins carefully selected to match exactly, and cured and prepared in the same way as other more famous furs. They have only just been sewn together, and the rug is now spread on the sofa to dry. She has made rugs, she will tell you, entirely of black cat skins, and very handsome they looked ; but not equal to this, which is wholly of the tabby. Certainly the gloss and stripe, the soft warmth and feel to the hand, seem to rival many foreign and costly importations. Besides carriage-rugs, the game-keeper's wife has made others for the feet—some many-coloured, like Joseph's coat.

All the cats to which these skins belonged were shot or caught in the traps set for vermin by her husband and his assistants. The majority were wild—that is, had taken up their residence in the woods, reverting to their natural state, and causing great havoc among the game. Feasting like this and in the joys of freedom, many had grown to a truly enormous size, not in fat, as the domestic animal does, but in length of back and limb. These afforded the best skins ; perhaps out of eight or nine killed but two would be available or worth preserving.

This gives an idea of the extraordinary number of cats

which stray abroad and get their living by poaching. They
invariably gravitate towards the woods. The instance in
point is taken from an outlying district far from a town,
where the nuisance is comparatively small ; but in the
preserves say from ten to twenty miles round London the
cats thus killed must be counted by thousands. Families
change their homes, the cat is driven away by the new
comer and takes to the field. In one little copse not more
than two acres in extent, and about twelve miles from
Hyde Park Corner, fifteen cats were shot in six weeks,
and nearly all in one spot—their favourite haunt. When
two or three wild or homeless animals take up their abode
in a wood, they speedily attract half a dozen hitherto tame
ones ; and, if they were not destroyed, it would be impossible to keep either game or rabbits.

She has her own receipts for preserving furs and
feathers, and long practice has rendered her an adept.
Here are squirrels' skins also prepared ; some with the
bushy tail attached, and some without. They vary in size
and the colour of the tail, which is often nearly white, in
others more deeply tinged with red. The fur is used to
line cloaks, and the tail is sometimes placed in ladies' hats.
Now and then she gets a badger-skin, which old country
folk used to have made into waistcoats, said to form an
efficacious protection for weak chests. She has made rugs
of several sewn together, but not often.

In the store-room upstairs there are a few splendid fox-skins, some with the tails tipped with white, others tipped with black. These are used for ladies' muffs, and look very handsome; the tail being occasionally curled round the muff. This sounds a delicate matter, and dangerously near the deadly sin of vulpecide. But it is not so. In these extensive woods, with their broad fringes of furze and heath, the foxes now and then become inconveniently numerous, and even cub-hunting will not kill them off sufficiently, especially if a great 'head' of game is kept up, for it attracts every species of beast of prey.

Besides the damage to game, the concentration of too many foxes in one district is opposed to the interest of the hunt—first, because the attendant destruction of neighbouring poultry causes an unpleasant feeling; next, because when the meet takes place the plethora of foxes spoils the sport. The day is wasted in 'chopping' them at every corner; the pack breaks up into several sections, despite whip, horn, and voice; and a good run across country cannot be obtained. So that once now and then a judicious thinning-out is necessary; and this is how the skins come into the hands of the keeper's wife. The heads go to ornament halls and staircases; so do the pads and occasionally the brush. The teeth make studs, set in gold; and no part of Reynard is thrown away, since the dogs eagerly snap up his body.

Once or twice she has made a moleskin waistcoat for a gentleman. This is a very tedious operation. Each

little skin has to be separately prepared, and when finished
hardly covers two square inches of surface.   Consequently
it requires several scores of skins, and the work is a year
or more about.   There is then the sewing together, which
is not to be accomplished without much patience and skill.
The fur is beautifully soft and glossy, with more resemblance to velvet than is possessed by any other natural
substance, and very warm.   Mittens for the wrists are also
made of it, and skull-caps.   Moleskin waistcoats used to
be thought a good deal of, but are now only met with
occasionally as a curiosity.

The old wooden mole-trap is now almost extinct,
superseded by the modern iron one, which anybody can
set up.   The ancient contrivance, a cylinder of wood, could
only be placed in position by a practised hand, and from
his experience in this the mole-catcher—locally called
' oont-catcher '—used to be an important personage in his
way.   He is now fast becoming extinct also—that is, as a
distinct handicraftsman spending his whole time in such
trapping.   He was not unfrequently a man who had once
occupied a subordinate place under a keeper, and when
grown too feeble for harder labour, supported himself in
this manner : contracting with the farmers to clear their
fields by the season.

Neither stoats' nor weasels' skins are preserved, except
now and then for stuffing to put under a glass case, though
the stoat is closely allied to the genuine ermine.   Polecats,

THE OLD MOLE-CATCHER.

too, are sometimes saved for the same purpose; in many
woods they seem now quite extinct. The otter skin is
valuable, but does not often come under the care of the
keeper's wife. The keeper now and then shoots a grebe
in the mere where the streamlet widens out into a small
lake, which again is bordered by water-meadows. This
bird is uncommon, but not altogether rare; sometimes two
or three are killed in the year in this southern inland
haunt. He also shoots her some jays, whose wings—as

likewise the black-and-white magpie—are used for the
same decorative purposes.   Certain feathers from the jay
are sought by the gentlemen who visit the great house, to
make artificial flies for salmon-fishing.   Of kingfishers she
preserves a considerable number for ladies' hats, and some
for glass cases.   Once or twice she has been asked to pre-
pare the woodpecker, whose plumage and harsh cry entitle
him to the position of the parrot of our woods.   Gentlemen interested in natural history often commission her
husband to get them specimens of rare birds ; and in the
end he generally succeeds, though a long time may elapse
before they cross his path.   For them she has prepared
some of the rare owls and hawks.   She has a store of pea-
cocks' feathers—every now and then people, especially
ladies, call at the cottage and purchase these things.
Country housewives still use the hare's 'pad' for several
domestic purposes—was not the hare's foot once kept in
the printing-offices ?

The keeper's wife has nothing to do with rabbits, but
knows that their skins and fur are still bought in large
quantities.   She has heard that geese were once kept in
large flocks almost entirely for their feathers, which were
plucked twice a year, she thinks ; but this is not practised
now, at least not in the south.   She has had snakes'
skins, or more properly sloughs, for the curious.   It is
very difficult to get one entire ; they are fragile, and so
twisted in the grass where the snake leaves them as to be

generally broken. Some country folk put them in their hats to cure headache, which is a very old superstition; but more in sport than earnest. There are no deer now in the park. There used to be a hundred years ago, and her husband has found several cast antlers in the wood.

ANTLERS ON THE STAIRCASE.

The best are up at the great house, but there is one on her staircase. Will I take a few chestnuts? It is winter— the proper time—and these are remarkably fine. No tree is apparently so capricious in its yield as the chestnut in English woods: the fruit of many is so small as to be worthless, or else it does not reach maturity. But these

large ones are from a tree which bears a fine nut: her
husband has them saved every year.   Here also are half
a dozen truffles if I will accept them : most that are found
go up to the great house ; but of late years they have not
been sought for so carefully, because coming in quantities
from abroad.   These truffles are found, she believes, in the
woods where the soil is chalky.   She used to gather many
native herbs ; but tastes have changed, and new seasonings
and sauces have come into fashion.

Out of doors in his work the assistant upon whom the
gamekeeper places his chief reliance is his own son—a lad
hardly taller than the gun he carries, but much older than
would be supposed at first sight.

It is a curious physiological fact that although open-
air life is so favourable to health, yet it has the apparent
effect of stunting growth in early youth. Let two
children be brought up together, one made to 'rough' it
out of doors, and the other carefully tended and kept
within ; other things being equal, the boy of the drawing-
room will be taller and to all appearance more developed
than his companion.   The labourers' children, for instance,
who play in the lonely country roads and fields all day,
whose parents lock their cottage doors when leaving for
work in the morning so that their offspring shall not gain
entrance and get into mischief, are almost invariably short
for their age.   In their case something may be justly
attributed to coarse and scanty food ; but the children of

working farmers exhibit the same peculiarity, and although
their food is not luxurious in quality, it is certainly not
stinted in quantity. Some of the ploughboys and carters'
lads seem scarcely fit to be put in charge of the huge cart-
horses who obey their shouted orders, their heads being

SMALL BOYS AND GREAT HORSES.

but a little way above the shafts—mere infants to look at.
Yet they are fourteen or fifteen years of age. With these,
and with the sons of farmers who in like manner work in
the field, the period of development comes later than with
town-bred boys. After sixteen or eighteen, after years of

hesitation as it were, they suddenly shoot up, and become great, hulking, broad fellows, possessed of immense strength. So the keeper's boy is really much more a man than he appears, both in years and knowledge—meaning thereby that local intelligence, technical ability, and unwritten education which is the resultant of early practice and is quite distinct from book learning.

From his father he has imbibed the spirit of the woods and all the minutiæ of his art. First he learned to shoot; his highest ambition being satisfied in the beginning when permitted to carry the double-barrel home across the meadow. Then he was allowed occasionally to fire off the charges left in after the day's work, before the gun was hung against the beam. Next, from behind the fallen trunk of an oak he took aim at a sitting rabbit which had raised himself on his hind-quarters to listen suspiciously—resting the heavy barrels on the tree, and made nervous by the whispered instructions from the keeper kneeling on the grass out of sight behind, 'Aim at his shoulder, lad, if he be sitting sidelong ; if a' be got his back to 'ee aim at his poll.' From this it was but a short step to be trusted with the single-barrel, and finally with the double ; ultimately having one of his own and walking his own distinct rounds.

He is now a keen shot, even better than his father ; for it is often observed that at a certain age young beginners in most manual arts reach an excellence which

in later years fails them. Perhaps the muscles are more
elastic, and respond instantaneously to the eye. This
mere boy at snap-shooting in the 'rough' will beat crack
sportsmen hollow. At the trap with pigeons he would
probably fail ; but in a narrow lane where the rabbits,
driven out by the ferrets, just pop across barely a yard of
open ground, where even a good shot may miss repeatedly,
he is 'death' itself to the 'bunnies.' So, too, with a wood-
hare—*i.e.* those hares that always lie in the woods as
others do in the open fields and on the uplands. They
are difficult to kill. They slip quietly out from the form
in the rough grass under the ashstole, and all you have for
guidance is the rustling and, perhaps, the tips of the ears,
the body hidden by the tangled dead ferns and 'rowetty'
stuff. When you try to aim, the barrel knocks against the
ashpoles, which are inconveniently near together, or the
branches get in the way, and the hare dodges round a tree,
and your cartridge simply barks a bow and cuts a tall
dead thistle in twain. But the keeper's lad, who had
waited for your fire, instantly follows, as it seems hardly
lifting his gun to his shoulder, and the hare is stopped by
the shot.

Rabbit-shooting, also, in an ash wood like this is trying to the temper ; they double and dodge, and if you
wait, thinking that the brown rascals must presently cross
the partially open space yonder, lo! just at the very edge
up go their white tails and they dive into the bowels of

D

the earth, having made for hidden burrows.   There is, of
course, after all, nothing but a knack in these things.
Still it is something to have acquired the knack.   The

THE KEEPER'S SON SHOOTING LEFT-HANDED.

lad, if you ask him, will proudly show off several gun-
tricks, as shooting left-handed, placing the butt at the left
instead of the right shoulder and pulling the trigger with

the left finger. He will knock over a running rabbit like this ; and at short distances can shoot with tolerable certainty from under the arm without coming to the 'present,' or even holding the gun out like a pistol with one hand.

By slow degrees he has obtained an intimate acquaintance with every field on the place, and no little knowledge of natural history. He will decide at once, as if by a kind of instinct, where any particular bird or animal will be found at that hour.

He is more bitter than his father against poachers, and would like to see harder measures dealt out to them ; but his chief use is in watching or checking the assistants, who act as beaters, ferreters, or keep up the banks and fences about the preserves, etc. Without a doubt these men are very untrustworthy, and practise many tricks. For instance, when they are set to ferret a bank, what is to prevent them, if the coast is clear, from hiding half a dozen dead rabbits in a burrow ? Digging has frequently to be resorted to, and thus they can easily cast earth over and conceal the entrance to a hole. Many a wounded hare and pheasant that falls into the hands of the beaters never makes its appearance at the table of the sportsman ; and doubtless they help themselves to the game captured in many a poacher's wire before giving notice of the discovery to the head man.

Some of these assistants wear waistcoats of calfskin

with the hair on it. The hair is outside, and the roan-and-white colour has a curious appearance: the material is said to be very warm and durable. Such waistcoats were common years ago; but of late the looms and spindles of the manufacturing districts have reduced the most outlying of the provinces to a nearly dead uniformity of shoddy.

One pair of eyes cannot be everywhere at once; consequently the keeper, as his son grew up, found him a great help in this way: while he goes one road the lad goes the other, and the undermen never feel certain that some one is not about. Perhaps partly for this reason the lad is not a favourite in the village, and few if any of the other boys make friends with him. He is too loyal to permit of their playing trespass—he looks down on them as a little lower in the scale. Do they ever speak, even in the humblest way, to the proprietor of the place? In their turn they ostracise him after their fashion; so he becomes a silent, solitary youth, self-reliant, and old for his years.

He is a daring climber: as after the hawk's nest, generally made in the highest elms or pines—if that species of tree is to be found—taking the young birds to some farmhouse where the children delight in living creatures. Some who are not children, or are children of 'a larger growth,' like to have a tame hawk in the garden, clipping the wings so that it shall not get away. Hawks

have most amusing tricks, and in time become comparatively tame, at least to the person who feeds them. The
beauty of the hawk's eye can hardly be surpassed: full,
liquid, and piercing. In this way the keeper's boy often

OWLS IN EVERY BARN.

gets a stray shilling; also for young owls, which are still
kept in some country houses, in the sheds or barns, to
destroy the mice. When the corn was threshed with the
flail, and was consequently exposed to the ravages of these
creatures (if undisturbed they multiply in such numbers as
would scarcely be credited), owls were almost domestic

birds, being domiciled in every barn.   Now they are more
objects of curiosity, though still useful when large teams
of horses are kept and require grain.

The keeper's boy sells, too, young squirrels from time
to time, and the eggs of the rarer birds.   In short, he has
imbibed all the ways of the woods, and is an adept at
everything, from 'harling' a rabbit upwards.   By-the-bye,
what is the etymology of 'harling,' which seems to have
the sense of entangling ?   It is done by passing the blade
of the knife between the bone of the thigh and the great
sinew—where there is nothing but skin—and then thrusting the other foot through the hole thus made.   The
rabbit or hare can then be conveniently carried by the
loop thus formed, or slung on a stick or the gun-barrel
across the shoulder.   Of course the 'harling' is not done
till the animal is dead.

The book-learning of the keeper's boy is rather limited,
for he was taught by the parish clerk and schoolmaster
before the Education Acts were formulated.   Still, he can
read, and pores over the weekly paper of rural sports, etc.,
taken for the guests at the great house and when out of
date sent down to the keeper's cottage.   In fact, he shows
a little too much interest in the turf columns to be quite
satisfactory to his father, who is somewhat anxious about
his acquaintance with the jockeys from the training-stables
on the downs hard by—an acquaintance he discourages
as tending to no good.   Like his father, he is never

seen abroad without a pair of leathern gaiters, and, if not a gun, a stout gnarled ground-ash stick in his hand.

The gamekeeper's calling naturally tends to perpetuate itself and become hereditary in his family. The life is full of attraction to boys—the gun alone is hardly to be resisted ; and, in addition, there are the animals and birds with which the office is associated, and the comparative freedom from restraint. Therefore one at least of his lads is sure to follow in his father's steps, and after a youth and early manhood spent out of doors in the woods it is next to impossible for him ever to quit the course he has taken. His children, again, must come within reach of similar influences, and thus for a lengthened period there must be a predisposition towards this special occupation.

Long service in one particular situation is not so common now as it used to be. Men move about from place to place, but wherever they are they still engage in the same capacity ; and once a gamekeeper always a gamekeeper is pretty nearly true. Even in the present day instances of families holding the office for more than one or two generations on the same estate may be found ; and years ago such was often the case. Occasionally the keeper's family has in this way, by the slow passage of time, become in a sense associated with that of his employer ; many years of faithful service sensibly abridging the social gulf between master and servant. The contrary holds equally true ; and so at the present day short terms

of service and constant changes are accompanied by a sharp
distinction separating employer and employé.

In such cases of long service the keeper holds a position more nearly resembling the retainer of the olden
time than perhaps any other 'institution' of modern life.
Pensioned off in his old age in the cottage where he was
born, or which, at any rate, he first entered as a child, he
potters about under his own vine and fig-tree—*i.e.* the
pear and damson trees he planted forty years before—
and is privileged now and then to give advice on matters
arising out of the estate.   He can watch the young broods
of pheasants still, and superintend the mixing of their
food : his trembling hand, upon the back of which the
corded sinews are so strongly marked now the tissue has
wasted, and over which the blue veins wander, can set a
trap when the vermin become too venturesome.

He is yet a terror to evil-doers, and in no jot abates
the dignity of more vigorous days ; so that the superannuated ancients whose task it is to sweep the fallen
leaves from the avenue and the walks near the great
house, or to weed the gravel drive in feeble acknowledgment of the charitable dole they receive, fall to briskly
when they see him coming with besom and rusty knife
wherewith to 'uck' out the springing grass.   He daily
gossips with the head gardener (nominal), as old or older
than himself; but his favourite haunt is a spot on the
edge of a fir plantation where lies a fallen 'stick' of timber.

Here, sheltered by the thick foliage of the fir and the hawthorn hedge at his back from the wind, he can sit on the log, and keep watch over a descending slope of

THE SUPERANNUATED KEEPER.

meadow bounding the preserves and crossed by footpaths, along which loiterers may come. His sturdy son now sways the sceptre of ash over the old woods, and other descendants are employed about the place.

Sometimes in the great house there may be seen the counterfeit presentment of such a retainer limned fifty years ago, with dog and gun, and characteristic background

of trees.   His wife has perhaps survived till recently—
strong and hale almost to the last ; the most voluble
gossip of the hamlet, full of traditions relating to the great
house and its owners ; a virago if crossed.   It is recorded
that upon one occasion in her prime she confronted a
couple of poachers, and, by dint of tongue and threats of
assistance close at hand, forced them to retire.   It was at
night that, her husband being from home and hearing shots
in the wood, she sallied forth armed with a gun, faced the
poachers, and actually drove them away, doubtless as much
from fear of recognition as of bodily injury, though even
that she was capable of inflicting, being totally fearless.

Nothing can be more natural than that when a man
has shown an earnest desire to give satisfaction and proved
himself honest and industrious, his employer should exhibit
an interest in the welfare of his family.   Now and then a
small farm may be found in the hands of a man descended
from or connected with a keeper.   To successfully work a
tenancy of such narrow limits it is necessary that the
occupier should himself labour in the field from morn till
dewy eve—the capacity to work being even more essential
than capital ; and so it happens that the smaller farms are
occasionally held by men who have risen from the lower
classes.   The sons of keepers also become gentlemen's
servants, as grooms, etc., in or out of the house.

A proposal was not long since made that gentlemen
who had met with misfortune or were unable to obtain

congenial employment, should take service as gamekeepers
—after the manner in which ladies were invited to become
'helps.' The idea does not appear to have received much
practical support, nor does it seem feasible, looking at the
altered relations of society in these days. A gentleman
'out of luck,' and with a taste for outdoor life and no
objection to work, could surely do far better in the
colonies, where he could shoot for his 'own hand,' and in
course of time achieve an independence, which he could
never hope to attain as a gamekeeper.

In the olden times, no doubt, younger brothers did
become, in fact, gamekeepers, head grooms, huntsmen, etc.,
to the head of the family. There was less of the sense of
servitude and loss of dignity when the feeling of clanship
was prevalent, when the great house was regarded as the
natural and proper resource of every cadet of the family.
But all this is changed. And for a man of education to
descend to trapping vermin, filling cartridges, and feeding
pheasants all his life, would be a palpable absurdity with
Australia open to him and the virgin soil of Central Africa
eager for tillage.

Neither is every man's constitution capable of with-
standing the wear and tear of a keeper's life. I have deline-
ated the more favourable side already; but it has its shadows.
Robust health, power of bearing fatigue, and above all of
sustaining constant exposure in our most variable climate,
are essential. No labourer is so exposed as the keeper:

the labourer does not work in continued wet, and he is sure
of his night's rest.   The keeper is often about the best part
of the night, and he cannot stay indoors because it rains.

The woods are lovely in the sunshine of summer ; they
are full of charm when the leaves are bursting forth in
spring or turning brown with the early autumn frosts ; but
in wet weather in the winter they are the most wretched
places conceivable in which to stroll about.   The dead
fern and the long grass are soaked with rain, and cling
round the ankles with depressing tenacity.   Every now
and then the feet sink into soddened masses of decaying
leaves—a good deal, too, of the soil itself is soft and peaty,
being formed from the decomposed vegetation of years ;
while the boughs against which the passer-by must push
fly back and send a cold shower down the neck.   In fog
as well as in rain the trees drip continuously ; the boughs
condense the mist and it falls in large drops—a puff of
wind brings down a tropical shower.

In warm moist weather the damp steam that floats in
the atmosphere is the reverse of pleasant.   But a thaw is
the worst of all, when the snow congealed on the branches
and against the trunks on the windward side, slips and
comes down in slushy, icy fragments, and the south-west
or south-east wind, laden with chilling moisture, penetrates
to the very marrow.   Even Robin Hood is recorded to
have said that he could stand all kinds of weather with
impunity, except the wind which accompanies a thaw.

Wet grass has a special faculty for saturating leather. The very boots with which you may wade into a stream up to your ankles in perfect comfort are powerless to keep out the dew or raindrops on the grass-blades. The path of the keeper is by no means always strewn with flowers.

Probably the number of keepers has much increased of recent years, since the floodtide of commercial prosperity set in. Every successful merchant naturally purchases an estate in the country, and as naturally desires to see some game upon it. This necessitates a keeper and his staff. Then game itself—meaning live game—has become a marketable commodity, bought and sold very much as one might buy a standing crop of wheat.

Owners of land, whose properties are hardly extensive enough to enable them to live in the state which is understood by the expression 'country seat,' frequently now resort to certain expedients to increase their incomes. They maintain a head of game large in comparison with the acreage : of course this must be attended to by a resident keeper; and they add to the original mansion various attractive extra buildings—*i.e.* a billiard-room, conservatories, and a range of modern stabling. The object, of course, is to let the house, the home farm, and the shooting for the season ; including facilities for following the hunt. The proprietor is consequently only at home in the latter part of the spring and in the summer—sometimes not even then.

Again, there are large properties, copyhold, or held under long leases from corporate bodies, the tenants having the right to shoot.   Instead of exercising the power themselves, they let the shooting.   It consists mainly of partridges, hares, and rabbits; and one of their men looks after the game, combining the keeping a general watch with other duties.   Professional men and gentlemen of independent income residing in county towns frequently take shooting of this kind.   The farmers who farm their own land often make money of their game in the same way.

Gentlemen, too, combine and lease the shooting over wide areas, and of course find it necessary to employ keepers to look after their interests.   The upper class of tradesmen in county and provincial towns where any facilities exist now sometimes form a private club or party and rent the shooting over several farms, having a joint-stock interest in one or more keepers.   Poor land which used to be of very little value has, by the planting of covers and copses, and the erection of a cottage for the keeper and a small 'box' for temporary occupation, in many cases been found to pay well if easily accessible from towns.   Game, in short, was never so much sought after as at present; and the profession of gamekeeping is in no danger of falling into decay from lack of demand for the skill in woodcraft it implies.

PHEASANTS.

CHAPTER III.

In the Fields.

MUCH other work besides preventing poaching falls upon
the keeper, such as arranging for the battue, stopping fox
'earths' when the hounds are coming, feeding the young
birds and often the old stock in severe weather, and even
some labour of an agricultural character.

A successful battue requires no little *finesse* and patience exercised beforehand ; weeks are spent in preparing for the amusement of a few hours. The pheasants are sometimes accustomed to leave the wood in a certain direction chosen as specially favourable for the sport— some copses at a little distance are used as feeding-places, so that the birds naturally work that way. Much care is necessary to keep a good head of game together, not too much scattered about on the day fixed upon. The difficulty is to prevent them from wandering off in the early morning ; and men are stationed like sentinels at the usual points of egress to drive them back. The beaters are usually men who have previously been employed in the woods and possess local knowledge of the ground, and are instructed in their duties long before : nothing must be left to the spur of the moment. Something of the skill of the general is wanted to organise a great battue : an instinctive insight into the best places to plant the guns, while the whole body of sportsmen, beaters, keepers with ammunition, should move in concert.

The gamekeeper finds his work fall upon him harder now than it used to do : first, sportsmen look for a heavier return of killed and wounded ; next, they are seldom willing to take much personal trouble to find the game, but like it in a manner brought to them ; and, lastly, he thinks the shooting season has grown shorter. Gentlemen used to reside at home the greater part of the winter, and spread their shooting over many months.   Now, the seaside season has moved on, and numbers are by the beach at the time when formerly they were in the woods.   Then others go abroad ; the country houses now advertised as 'to let' are almost innumerable.   Time was when the local squire would have thought it derogatory to his dignity to make a commodity of his ancient mansion; now there seems quite a competition to let, and absenteeism is a reality of English as well as Irish country life.   At least, such is the gamekeeper's idea, and he finds a confirmation of it in the sudden rush, as it were, made upon his preserves.   Gentlemen who once spent weeks at the great house, and were out with him every day till he grew to understand the special kind of sport which pleased them most, and could consequently give them satisfaction, are now hardly arrived before they are gone again.   With all his desire to find them game he is often puzzled, for game has its whims and fancies, and will not accommodate itself to their convenience.

Then the keeper thinks that shooting does not begin so early as it once did.   Partridges may be found in the market on the morning of the glorious First of September ; but if you ask him how they get there, your reply is a nod and a wink.   Nobody gets up early enough in the morning for that now : very often the first day passes by without a single shot being fired.   The eagerness for the stubble and its joys is not so marked.   This last season the late harvest interfered very much with shooting ; you

E

cannot walk through wheat or barley, and while the crops
are standing the partridges have too much cover.

Many gentlemen, again, keep their pheasants till nearly
Christmas : October goes by frequently without a bird
being brought down in some preserves.   Early in the new
year, if the weather be mild, as it has been so often latterly,
the birds begin to show signs of a disposition to pair off,
and in consequence the guns are laid aside before the cer-
tificates expire.   So that the keeper thinks the actual
shooting season has grown shorter and the sport is more
concentrated, and taken in rushes, as it were.   This causes
additional work and anxiety.   If the family are away
they still require a regular and sometimes a large supply
of game for the table, which he has to keep up himself—
assistants could hardly be trusted : the opportunity is too
tempting.

Though a loyal and conscientious man, in his secret
heart he does not like the hounds : and though of course
he gets tipped for stopping the earths, yet it is a labour
not exactly to his taste.   The essence of game-preserving
is quiet, repose ; the characteristic of the hunt is noise,
horn, whoop, whip, the cry of the hounds, and the crash of
the bushes as the field takes a jump.   Students and book-
worms like the quiet dust which settles in their favourite
haunts—the housemaid's broom is fatal to retrospective
thought : so the gamekeeper views the squadrons charging
through his cherished copses, 'poaching' up the green-

sward of the winding 'drives,' breaking down the fences, much as the artist views the sacrilegious broom 'putting his place to rights.' Pheasant, and hare, and rabbit all are sent helter-skelter anywhere, and take a day or two to settle down again.

Yet it is not so much the real genuine hunt that he dislikes: it is the loafers it brings together on foot. Roughs from the towns, idle fellows from the villages, cobblers, tinkers, gipsies, the nondescript 'residuum,' all congregate in crowds, delighted at the chance of penetrating into the secret recesses of woods only thrown open two or three times a year. It is impossible to stay the inroad—the gates are wide open, the rails pulled down, and trespass is but a fiction for the hour. To see these gentry roaming at their ease in his woods is a bitter trial to the keeper, who grinds his teeth in silence as they pass him with a grin, perfectly aware of and enjoying his spleen. Somehow or other these fellows always manage to get in the way just where the fox was on the point of breaking cover; if he makes a clear start and heads for the meadows, before he has passed the first field a ragged jacket appears over the hedge, and then the language of the huntsman is not always good to listen to.

The work of rearing the young broods of pheasants is a trying and tedious one. The keeper has his own specific treatment, in which he has implicit faith, and laughs to scorn the pheasant-meals and feeding-stuffs

advertised in the papers.   He mixes it himself, and likes
no one prying about to espy his secret, though in reality
his success is due to watchful care and not to any
particular nostrum.   The most favourable spot for rearing

TENDING THE YOUNG BIRDS.

is a small level meadow, if possible without furrows, which
has been fed off close to the ground and is situated high
and dry, and yet well sheltered with wood all round.
Damp is a great enemy of the brood, and long grass wet
with dew in the early morning sometimes proves fatal if

the delicate young birds are allowed to drag themselves through it.

Besides the coops, here and there bushes, cut for the purpose, are piled in tolerably large heaps. The use of these is for the broods to run under if a hawk appears in the sky ; and it is amusing to watch how soon the little creatures learn to appreciate this shelter. In the spring the greater part of the keeper's time is occupied in this way : he spends hours upon hours in the hundred and one minutiæ which ensure success. This breeding-time is *the* great anxiety of the year : on it all the shooting depends. He shakes his head if you hint that perhaps it would save trouble to purchase the pheasants ready for shooting from the dealers who now make a business of supplying them for the battue. He looks upon such a practice as the ruin of all true woodcraft, and a proof of the decay of the present generation.

In addition to the pheasants, the partridges, wild as they are, require some attention—the eggs have to be looked after. The mowers in the meadows frequently lay their nests bare beneath the sweep of the scythe : the old bird sometimes sits so close as to have her legs cut off by the sharp steel. Occasionally a rabbit, in the same way, is killed by the point of the blade as he lingers in his form. The mowers receive a small sum for every egg they bring, the eggs being placed under brood hens, kept for the purpose. But as a partridge's egg from one field is precisely like one from another field, the keeper may find, if he does not look pretty sharp after the mowers on the estate, that they have been bribed by a trifle extra to carry the eggs to another man at a distance. A very unpleasant feeling often arises from suspicions of this kind.

His agricultural labour consists in superintending the cultivation of the small squares left for the growth of grain in the centre of the copses, to feed and attract the pheasants, and to keep them from wandering. These have to be dug up with the spade—there would be no room for using a plough—and spade-husbandry is rather slow work. An eye has therefore to be kept on the labourers thus employed lest they get into mischief. The grain (on the straw) is sometimes given to the birds laid across skeleton trestles, roughly made of stout ash sticks, so as to raise it above the ground and enable them to get at it better.

Ash woods are cut every year, or rather they are mapped out into so many squares, the poles in which come to maturity in succession—while one is down another is growing up, and thus in a fixed course of years the entire wood is thrown and renovated. A certain time has, of course, to be allowed for purchasers to remove their property, and, as the roads through the woods are often axle-deep in mud, in a wet spring it has frequently to be extended. So many men being about, the keeper has to be about also: and then, when at last the gates are

CATTLE IN THE COPSE.

nailed up, the cattle turned out to grass in the adjacent fields often break in and gnaw the young ash-shoots. In this way a trespassing herd will throw back acres of wood for a whole year, and destroy valuable produce. Properly speaking, this should come under the attention of the bailiff or steward of the demesne; but as the keeper and his men are so much more likely to discover the cattle first, they are expected to be on the watch.

After spending so many years of his life among trees, it is natural that the keeper should feel a special interest, almost an affection for them. A branch ruthlessly torn down, a piece of bark stripped from the trunk with no possible object save destruction, a nail driven in—perhaps to break the teeth of the saw when at last the tree comes to be cut up into planks—these things annoy him almost as much as if the living wood were human and could feel. For this reason, he too, like the members of the hunt, cordially detests the use of wire for fencing, now becoming so frequent. It cuts into the trees, and checks their growth and spoils their symmetry, if it does not actually kill them.

Sometimes the wire, which is stout and strong, is

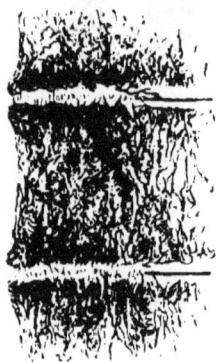

DIAGRAM TO SHOW DAM-
AGE DONE BY IRON
WIRE-FENCING.

twisted right round the stem of a young oak, say a foot or more in diameter, which is thus made to play the part of a post. A firmer support could not be found ; but as the tree swells with the rising sap, and expands year by year, the iron girdle circling about it does not 'give' or yield to this slow motion. It bites into the bark, which in time curls over, and so actually buries the metal in the growing wood. Now this cannot but be injurious to the tree itself, and it is certainly unsightly.

One wire is seldom thought enough. Two or three are stretched along, and each of these causes an ugly scar. If allowed to remain long enough, the young wood will solidify and harden about the wire, which then cannot be withdrawn ; and in consequence, when taken finally to the sawpit, some three or four feet of the very 'butt' and best part of the trunk will be found useless. No sawyer will risk his implement—which requires some hours' work to sharpen—in wood which he suspects to contain concealed iron. So that, besides the injury to the appearance of the tree, there is a pecuniary loss. Even when the wires are not twisted round, but merely rub against one side of the bark, the same scars are caused there, though not to such an extent. Rough and strong as the bark seems to the touch, it speedily abrades under the constant pressure of the metal.

The keeper thinks that all those owners of property who take a pleasure in their trees should see to this and prevent it. There is nothing so detestable as this wire-fencing in his idea. You cannot even sit upon it for rest, as you can on the old-fashioned post and rails. The convenient gaps which used to be found in every hedge at the corner are blocked now with an ugly rusty iron string stretched across, awkward to get over or under ; while as for a horseman getting by, you cannot pull it down as you could 'draw' a wooden rail, and if you try to uncoil it from the blackthorn stem to which it is attached, the jagged end is tolerably certain to scrape the skin from your fingers.

The keeper looks upon this simply as another sign of

the idleness and dislike of taking trouble characteristic of
the times.   To set up a line of posts and rails requires
some little skill ; a man must know his business to stop
a gap with a single rail or pole, fixing the ends firmly
in among the underwood ; even to fit thorn bushes in
properly, so as to effectually bar the way, needs some
judgment : but anybody can stretch a wire along and
twist it round a tree.   Hedge-carpentering was, in fact, a
distinct business, followed by one or two men in every
locality ; but iron now supplants everything, and the
hedges themselves are disappearing.

When the hedgers and ditchers were put to work to
cut a hedge—the turn of every hedge comes round once
in so many years—they used to be instructed, if they
came across a sapling oak, ash, or elm, to spare it, and
cut away the bushes to give it full play.   But now they
chop and slash away without remorse, and the young
forest-tree rising up with a promise of future beauty falls
before the billhook.   In time the full-grown oaks and
elms of the hedgerow decay, or are felled ; and in consequence of this careless destruction of the saplings there
is nothing to fill their place.   The charm of English
meadows consisted in no small degree in the stately trees,
whose shadows lengthened with the declining sun and
gave such pleasant shelter from the heat.   Soon, however,
if the rising generation of trees is thus cut down, they
must become bare, open, and unlovely.

There is another mistake, often committed by owners of timber, who go to the other extreme, and in their intense admiration of trees refuse to permit the felling of a single one. Now in the forest or the woodlands, away from the park or pleasure-grounds, the old hollow trees are things of beauty, and to cut them down for firewood seems an act of vandalism. But it is quite another thing with an avenue or those groups which dot the surface of a park. Here, if a tree falls and there is no other to take its place, a gap is the result, which cannot be filled up, perhaps, under fifty or sixty years.

Let any one stroll along beneath a stately avenue of elm or beech, such as are not difficult to find in rural districts, and are the pride and boast—and justly so—of this country, and, examining the trees with critical eye, what will he see? Three or four elms, I will say, are passed, and are evidently sound ; but the fifth—a careless observer might go by it without remarking anything un-usual—is really rotten to the centre. At the foot of the huge trunk, and growing out of it, is a bunch of sickly-looking fungi. Thrust your walking-stick sharply against the black wood there and it penetrates easily, and with a little pushing goes in a surprising distance ; the tree seems undermined with rottenness. This decay really runs up the trunk perpendicularly : look, there are signs of it above at the knot-hole, thirty feet high, where more fungus is flourishing, as it always does in dead damp

· wood. The rain soaks in there, and filtrates slowly down the trunk, whose very heart as it were is eaten away, while outside all is fair enough.

Presently there arises a mighty wind, the tree snaps clean off twenty feet above the ground, and the upper part falls, a ponderous ruin, carrying with it one of the finest boughs of its nearest companion, and destroying its symmetry also. When examined, it appears that the trunk is totally useless as timber : this noble-seeming elm is fit for nothing but fuel. Or, perhaps, if there be water meadows on the estate, the farmers may be glad of it to act as a huge pipe to convey the fertilising stream across a ditch, or over a brook lying at a lower level. For this purpose, of course, the rotten part is scooped out : often the trunk is sawn down the middle, so as to make a double length.

But what a gap it has left in the great avenue ! In a minute the growth of a century gone, the delight of generations swept away, and no living man, hardly the heir in his cradle, can hope to see that unsightly gap filled up.

The keeper does not hesitate to say that of the great trees in the avenues numbers stand in constant danger of such overthrow ; and so it is that by slow degrees so many of the kings of the forest have disappeared without leaving successors. No care is taken to plant fresh saplings, no care is taken to select and remove the trees which have passed the meridian of their existence, and the final result is the extinction of the avenue or group. Perhaps the temper of the times is to blame for this neglect : men look only to the day and live fast. There is a sense of uncertainty in the atmosphere of the age : no one can be sure that the acorns he plants will be permitted to reach their prime ; the hoofs of the 'iron horse' may trample them down as fresh populations grow. So the avenues die out, and the keeper mourns to think that in the days to come their place will be vacant.

Suddenly he pauses in his walk, stoops, and points out to me in the grass the white, smooth, round knob-like tops of several young mushrooms which are pushing their way up. He carefully covers these with some pieces of dead bark and desiccated dung, so that none of 'them lurching fellows as comes round shan't see 'em '—with a wink at his own cunning—so as to preserve them till they have grown larger. He advises me never to partake of mushrooms unless certain that they have not grown under oak trees : he will have it that even the true edible mushroom is hurtful if it springs beneath the shadow of the oak. And he is not singular in this belief.

Chatting about trees, he points out one or two oaks, not at all rotten, but split half-way up the trunk—the split is perfectly visible—yet they have not been struck by lightning ; and he cannot explain it. Looking back upon the wood as we leave it with intense pride in his trees, he gives me a rough version of the old story : how a knight of ancient days, who had done the king some great service was rewarded with a broad tract of land which he was to hold for three crops.   He sowed acorns, and thus secured himself and descendants a tenure of almost 3000 years, at least, according to Dryden :—

> The monarch oak, the patriarch of trees,
> Shoots, rising up, and spreads by slow degrees ;
> Three centuries he grows, and three he stays
> Supreme in state, and in three more decays.

The keeper wishes he had such an opportunity.   The knight, in his idea, reached the acme of wisdom with his three crops of nearly a thousand years each.

His own kingdom may be said to begin with the park, and the land ' in demesne,' to quote the quaint language of the Domesday Book : a record not without its value as an outline picture of English scenery eight centuries ago ; telling us that near this village was a wood, near that a stretch of meadow and a mill, here again arable land and corn waving in the breeze, and everywhere the park and domain of the feudal lords.   The beauty of the park con-sists in its ' breadth ' as an artist would say—the meadows with their green frames of hedges are cabinet pictures, lovely, but small ; this is life size, a broad cartoon from the hand of Nature.   The sward rises and rolls along in un-dulations like the slow heave of an ocean wave.   Besides the elms there is a noble avenue of limes, and great oaks scattered here and there, under whose ample shade the cattle repose in the heat of the day.

THE PARK.

In summer from out the leafy chambers of the limes there falls the pleasant sound of bees innumerable, the voice of whose trembling wings lulls the listening ear as the drowsy sunshine weighs the eyelid till I walk the avenue in a dream. It leads out into the park—no formal gravel drive, simply a footpath on the sward between the flowering trees : a path that becomes less and less marked as I advance, and finally fades away, where the limes cease, in the broad level of the opening 'greeny field.' These

honey-bees seem to fly higher and to exhibit much more
activity than the great humble-bee : here in the limes they
must be thirty feet above the ground. Wasps also frequently wing their way at a considerable elevation, and
thus it is that the hive-bee and the wasp so commonly
enter the upper windows of houses. When its load of
honey is completed, the bee, too, returns home in a nearly
straight line, high enough in the air to pass over hedges
and such obstacles without the labour of rising up and
sinking again.

The heavy humble-bee is generally seen close to the
earth, and often goes down into the depths of the dry
ditches, and may there be heard buzzing slowly along
under the arch of brier and bramble. He seems to lose
his way now and then in the tangled undergrowth of the
woods ; and if a footstep disturbs and alarms him it is
amusing to see his desperate efforts to free himself hastily
from the interlacing grass-blades and ferns.

When the sap is rising, the bark of the smaller shoots
of the lime-tree ' slips' easily—*i.e.* it can be peeled in
hollow cylinders if judiciously tapped and loosened by
gentle blows from the back of a knife. The ploughboys
know this, and make whistles out of such branches, as they
do also from the willow, and even the sycamore in the
season when the sap comes up in its floodtide.

It is difficult to decide at what time of the year the
park is in its glory. The may-flower on the great haw-

thorn trees in spring may perhaps claim the pre-eminence, filling the soft breeze with exquisite odour. These here are trees, not bushes, standing separate, with thick gnarled stems so polished by the constant rubbing of cattle as almost to shine like varnish. The may-bloom, pure white in its full splendour, takes a dull reddish tinge as it fades, when a sudden shake will bring it down in showers. A flowering tree, I fancy, looks best when apart and not one of a row. In the latter case you can only see two sides and not all round it. Here tall horse-chestnut trees stand single—one great silvery candelabrum of blossom. Wood-pigeons appear to have a liking for this tree. Nor must the humble crab-tree be forgotten ; a crab-tree in bloom is a lovely sight.

The idea of a park is associated with peace and pleasure, yet even here there is one spot where the passions of men have left their mark. As previously hinted, the gamekeeper, like most persons with little book-learning and who take their impressions from nature, is somewhat superstitious, and regards this place as ' unkid '—*i.e.* weird, uncanny. One particular green 'drive' into the wood opening on the park had always been believed to be a part of a military track used many ages ago, but long since ploughed up for the greater part of its length, and only preserved here by the accident of passing through a wood. At last some labourers grubbing trees near the mouth of the drive came upon a number of human

F

skeletons, close beneath the surface, and in their confused arrangement presenting every sign of hasty interment, as if after battle. Since then the keeper avoids the spot ; nor will he, hardy as he is, go near it at night ; not even in the summer moonlight, when the night is merely a prolongation of the day.

There is nothing unusual in such a discovery : skeletons are found in all manner of places. I recollect seeing one dug out from the bank of a brook within two feet of the stream. The place was perhaps in the olden time covered with forest (traces of forest are to be found everywhere, as in the names of hamlets), and therefore more concealed than at present. Or, possibly, the stream, in the slow passage of centuries, may have worn its way far from its original bed.

It is strange to think of, yet it is true enough, that, beautiful as the country is, with its green meadows and graceful trees, its streams and forests and peaceful homesteads, it would be difficult to find an acre of ground that has not been stained with blood. A melancholy reflection this, that carries the mind backwards, while the thrush sings on the bough, through the nameless skirmishes of the Civil War, the cruel assassinations of the rival Roses, down to the axes of the Saxons and the ghastly wounds they made. Everywhere under the flowers are the dead.

Not this park in particular, but others as well form pages of history. The keeper, in fact, can claim an ancient origin for his office, dating back to the forester with a 'mark' a year and a suit of green as his wages, and numbering in his predecessors Joscelin, the typical keeper in Scott's novel of 'Woodstock,' who aided the escape of King Charles. Ever since the days of the Norman king who loved the tall deer as if he were their father—in the words of his contemporary—and set store by the hares that they too should go free, the keeper has not ceased out of the land.

There are always more small birds at the edge or just outside a wood than inside it ; so that after leaving a meadow with blackbird, thrush, and finches merry in the hedges, the wood seems quite silent and deserted save by a solitary robin. This is speaking of the smaller birds. The great missel-thrush especially delights in the open space of the park dotted with groups of trees. The missel-thrush is a lonely bird, and somehow seems like an outlaw—as if, though not precisely dangerous, he was looked upon with suspicion by the other birds, which will frequently quit a bush or tree directly he alights upon it. Yet he builds near houses, and year after year in the same spot. I knew a large yew-tree which stood almost in front and within a few yards of a sitting-room window in which the missel-thrush had regularly built its nest for twelve successive years. These birds are singularly bold in defence of the nest, flying round and chattering at those who would disturb it.

In the ha-ha wall of the park, which is made of loose
stones or without mortar, the tomtit, or titmouse, has his
nest.   He creeps in between the stones, following the
crannies for a surprising distance.   Near here the partridges roost on the ground ; they like an open space far
from hedges, afraid, perhaps, of weasels and rats.   On the
other side, where the wood comes up, if you watch quietly,
the pheasants step in lordly pride out into the grass ; so
that there is no place without its especial class of life.

Perhaps, with the exception of our parks and hills,
there is scarcely any portion of southern England now
where a grand charge of cavalry could take place—scarcely
any open champaign country fit for operations of that kind
with horse.   In the Civil War even, how constantly we
read of 'lining the hedge with match,' and now with en-
closures everywhere the difficulty would apparently be
great, despite good roads.

Park-fed beef is thought by many to be superior,
because the cattle run free—almost wild—the entire year
through, winter and summer, and have nothing but their
natural food, grass and hay : in strong contrast with the
bullocks shut up in stalls and forced forward with artificial
food.   A great number of parks have been curtailed in
size as land became more valuable—the best ground being
selected and hedged off for purely agricultural purposes ;
so that it not uncommonly happens that the actual park is
the poorest soil in the district, having for that reason re-

mained longest in a condition nearly resembling the original state of the country.   So that when agitators of
Communistic views lay stress upon the waste of land used
for pleasure purposes they frequently declaim in utter
ignorance of the facts, which are in exact opposition to
their theories.

Like animals and birds, plants have their favourite
haunts : violets love a bank with a southern aspect, especially if there be a hedge at the back for further shelter.
Where you have by chance lighted upon a wild flower once
you may generally reckon upon finding it again next year
—such as the white variety of the bluebell or wild
hyacinth, for which, unless you mark the place, you may
search in vain amid the crowded blue bloom of the commoner sort.   The orchis, with its purple flower and dark
green spotted leaf, in the virtue of whose roots as a love-
potion the old people still believe, the strange-looking
adder's tongue, the modest wild strawberry, with its tiny
but piquant-flavoured fruit, all have their special resorts.
Even the cowslips have their ways : by brooks sometimes
a larger variety grows ; nor is there a sweeter flower than
its delicate yellow with small velvety brown spots, like
moles on beauty's cheek.

In autumn, when the leaves turn colour, the groups of
trees in the park are more effective in an artistic point of
view than those in the woods (unless overlooked from a
hill close by, when it is like glancing along a roof of gold),

because they stand out clear, and are not confused or lost in the general glow. But it is evening now ; and see— yonder the fox steals out from the cover, wending his way down into the meadows, where he will follow the furrows along their course, mousing as he goes.

CHAPTER IV.

His Dominions :—the Woods—Meadows—and Water.

THERE is a part of the wood where the bushes grow but thinly and the ashstoles are scattered at some distance from each other. It is on a steep slope—almost cliff—where the white chalk comes to the surface. On the edge above rise tall beech trees with smooth round trunks, whose roots push and project through the wall of chalk, and bend downwards, sometimes dislodging lumps of rubble to roll headlong among the bushes below. A few small firs cling half-way up, and a tangled mass of brier and bramble climbs nearly to them, with many a stout thistle flourishing vigorously.

To get up this cliff is a work of some little difficulty: it is done by planting the foot on the ledges of rubble, or in the holes which the rabbits have made, holding tight to roots which curl and twist in fantastic shapes, or to the woodbine hanging in festoons from branch to branch. The rubble under foot crumbles and slips, the roots tear up bodily from the thin soil, the branches bend, and the woodbine 'gives,' and the wayfarer may readily descend

much more rapidly than he desires. Not that serious con-
sequences would ensue from a roll down forty feet of slope;
but the bed of brier and bramble at the bottom is not so
soft as it might be. The rabbits seem quite at home upon
the steepest spot ; they may be found upon much higher
and more precipitous chalk cliffs than this, darting from
point to point with ease.

Once at the summit under the beeches, and there a
comfortable seat may be found upon the moss. The
wood stretches away beneath for more than a mile in
breadth, and beyond it winds the narrow mere glittering
in the rays of the early spring sunshine. The bloom is
on the blackthorn, but not yet on the may ; the hedges
are but just awakening from their long winter sleep, and
the trees have hardly put forth a sign. But the rooks are
busily engaged in the trees of the park, and away yonder
at the distant colony in the elms of the meadows.

The wood is restless with life : every minute a pigeon
rises, clattering his wings, and after him another ; and so
there is a constant fluttering and motion above the ashpoles. The number of wood-pigeons breeding here must
be immense. Later on, if you walk among the ash, you
may find a nest every half-dozen yards. It is formed of
a few twigs making a slender platform, on which the
glossy white egg is laid, and where the bird will sit till
you literally thrust her off her nest with your walking-
stick. Such slender platforms, if built in the hedgerow, so

soon as the breeze comes would assuredly be dashed to
pieces ; but here the wind only touches the tops of the
poles, and causes them to sway gently with a rattling
noise, and the frail nest is not injured.   When the pigeon

WOOD-PIGEON.

or dove builds in the more exposed hedgerows the nest is
stronger, and more twigs seem to be used, so that it is
heavier.

Boys steal these eggs by scores, yet it makes no
difference apparently to the endless numbers of these
birds, who fill the wood with their peculiar hoarse notes,

which some country people say resemble the words, 'Take two cows, Taffy.' The same good folk will have it that when the weather threatens rain the pigeon's note changes to 'Joe's toe bleeds, Betty.' The boys who steal the eggs have to swarm up the ashpoles for the purpose, and in so doing often stain their clothes with red marks. Upon the bark of the ash are innumerable little excrescences which when rubbed exude a small quantity of red juice.

The keeper detests this bird's-nesting; not that he cares much about the pigeons, but because his pheasants are frequently disturbed just at the season when he wishes them to enjoy perfect quiet. It is easy to tell from this post of vantage if any one be passing through the section of the wood within view, though they may be hidden by the boughs. The blackbirds utter a loud cry and scatter; the pigeons rise and wheel about; a pheasant gets up with a scream audible for a long distance, and goes with swift flight skimming away just above the ashpoles; a pair of jays jabber round the summit of a tall fir tree, and thus the intruder's course is made known. But the wind, though light, is still too cold and chilly as it sweeps between the beech trunks to remain at this elevation; it is warmer below in the wood.

At the foot of the cliff a natural hollow has been further scooped out by labour of man, and shaped into a small cave, large enough for three or four to sit in. It is partly supported by strong wooden pillars, and at the mouth a hut of slabs, thickly covered by furze-faggots, has been constructed, with a door, and with roof thatched with reeds from the lake. A rude bench runs round three sides; against the fourth some digging tools recline— strong spades and grub-axes for rooting out a lost ferret, left here temporarily for convenience. The place, rough as it is, gives shelter, and, throwing the door open, there is a vista among the ashpoles and the hazel bushes over-topped with great fir trees and more distant oaks. In the later spring this is a lovely spot, the ground all tinted with the shimmering colour of the bluebells, and the hazel musical with the voice of the nightingale.

Outside the wood, where the downland begins to rise gradually, there stretches a broad expanse of furze growing luxuriantly on the thin barren soil, and a mile or more in width. It has a beauty of its own when in full yellow blossom—a yellow sea of flower, scenting the air with an almost overpowering odour as of a coarser pineapple, and full of the drowsy hum of the bees busy in the interspersed thyme. It has another beauty later on when the thick undergrowth of heath is in bloom, and a pale purple carpet spreads around. Here rabbits breed and sport, and hares hide, and the curious furze-chats fly to and fro; and lastly, but not leastly, my lord Reynard the Fox loves to take his ease, till he finally meets his fate in the jaws of clamouring hounds, or is assassinated with the aid of 'villanous saltpetre.' He is not easily shot, and will stand a charge fired broadside at a short distance without the slightest injury or apparent notice, beyond a slight quickening of his pace. His thick fur and tough skin turn the pellets. Even when mortally wounded, life will linger for hours.

The ordinary idea of the fox is that of a flying frightened creature tearing away for bare existence ; he is really a bold and desperate animal. The keeper will tell you that once, when for some purpose he was walking up a deep dry ditch, his spaniel and retriever suddenly 'chopped' a fox, and got him at bay in a corner, when he turned, and in an instant laid the spaniel helpless and dying, and severely handled the retriever. Seeing his dogs so injured and the fox as it were under his feet, the keeper imprudently attempted to seize him, but could not retain his hold, and got the sharp white teeth clean through his hand.

Though but once actually bitten, he recollects being snapped at viciously by another fox, whom he found in broad daylight asleep in the hollow of a double mound, with scarcely any shelter, and within sixty yards of a house. Reynard was curled upon the ivy which in the hedges trails along the ground. The keeper crawled up on the bank and stopped, admiring the symmetry of the creature, when, purposely breaking a twig, the fox was up in a second, and snarled and snapped at his face, then slipped into the ditch and away. The fox is, in fact, quite as remarkable for boldness as for cunning. Last summer I met a fine fox on the turnpike road and close to a tollgate, in the middle of the day. He came at full speed with a young rabbit in his jaws, evidently but just captured, and did not perceive that he was observed till within twenty yards, when, with a single bound he cleared the sward beside the road, alighting with a crash in the bushes, carrying his prey with him.

Hares will sometimes, in like manner, come as it were to meet people on country roads. Is it that the eyes, being placed towards the side of the head, do not so readily catch sight of dangers in front as on the flanks, especially when the animal is absorbed in its purpose? Hares are peculiarly fond of limping at dusk along lonely roads.

Foxes, when they roam from the woods into the meadow-land, prefer to sleep during the day in those osier beds which are found in the narrow corners formed by the meanderings of the brooks. Between the willow-wands there shoots up a thick undergrowth of sedges, long coarse grass, and reeds; and in these the fox makes his bed, turning round and round till he has smoothed a place and trampled down the grass; then reclining, well sheltered from the wind. A dog will turn round and round in the same way before he lies down on the hearthrug.

These reeds sometimes grow to a great height, as much as ten or twelve feet. Along the Thames they are used, bound in bundles, to pitch the barges; when the hull has been roughly coated with pitch, one end of the bundle of reeds (thickest end preferred) is set on fire and passed over it to make it melt and run into the chinks. So, mayhap, the Saxon and Danish rovers may have used them to pitch the bottoms of their 'ceols' when worn from constantly grounding on the shallows and eyots.

Here in the furze too is the haunt of the badger. This animal becomes rarer year after year—the disuse of the great rabbit-warrens being one cause; still he lingers, and may be traced in the rabbit 'buries,' where he enlarges a hole for his habitation, sleeps during the day, and comes forth in the gloaming. In summer he digs up the wasps' nests, not, as has been supposed, for the honey, but for the white larvæ they contain: the wasp secretes no honey at all, and her nest is simply a series of cells in which the grubs mature. Some credit the fox with a fondness for the same food; and even the hornet's nest is said to be similarly ravaged. It is the nest of the humble-bee which the badger roots up for the honey. The humble-bee uses a tiny hole in a dry bank, sometimes a crack made by the heat in the earth, and really deposits true honey in the comb. It is very sweet, like that of the hive bee, but a little darker in colour and much less in quantity. The haymakers search for these

A BADGER AT HIS FRONT DOOR.

nests along the hedgerows in their dinner-hour, and eat the honey. There seem to be several sub-species of humble-bee, differing in size and habit. One has its nest as deep as possible in a hole ; another makes a nest with scarcely any protection beyond the thick moss of the bank, almost on the surface of the ground. The badger's hole has before it a huge quantity of sand, which he has thrown out, and upon which the imprint of his foot will be found, a mark, perhaps more like the spoor of the

large game of tropical forests than that left by any other
English animal. When seen it can ever afterwards be
instantly identified by the most careless observer.

In the meadows lower down, bounding the wood, the
hay is gone or is piled in summer ricks, which lean one
one way and another the other, and upon whose roofs,
sloping at an obtuse angle, the green snakes lie coiled in
the sunshine. Often when the waggon comes, and the
little rick is loaded, the 'pitch' of hay on the prong as it
is flung up carries with it a snake whirling in the air.
He falls on the sward and is instantly pounced upon by
the farmer's dog, who worries him, seizes him by the
middle and shakes him, while the snake twists and hisses
in vain. Some dogs will not touch snakes, others seem
to enjoy destroying them ; but it is noticeable that a dog
which previously has passed or avoided snakes, if once he
kills one, never passes another without slaughtering it.
A slime from the snake's skin froths over the dog's jaws,
and the sight is very unpleasant.

I have often tried to discover how the snakes get
upon these summer ricks.   Solomon could not understand
the ' way of a serpent upon the rock,' and the way of a
common snake up the summer rick seems almost as
inexplicable.   Though the roof or ' top ' is often very
much out of the proper conical shape, and sometimes
sinks down nearly to a level, the sides for a height of
three or four feet are generally perpendicular, affording no

projection of any kind whatever; hay is slippery, and the rick is, of course, too large for the snake to encircle it. Yet there they are commonly found, to the intense alarm of the labouring women, who never can get over their dislike of snakes, though they see them so frequently. The only way I can imagine by which they climb up is by means of the holes, or galleries, used by field-mice. In summer ricks there are sometimes many mice, and in pursuit of these the snake may find its way up through their 'runs.' Toads are also occasionally found on these ricks, and it is not exactly clear how they get there either; but their object is plain—i.e. the insects which swarm on the hay.

The thick hedgerows of these woodland meads are full of trees, and others stand out in groups in the grass, some of them hollow. Elms often become hollow, and so do oaks; the latter have such large cavities sometimes that one or more persons may easily crouch therein. This is speaking of an ordinary sized tree; there are many instances of patriarchs of the forest within whose capacious trunks a dozen might stand upright.

These hollow trees, according to woodcraft, ought to come down by the axe without further loss of time. Yet it is fortunate that we are not all of us, even in this prosaic age, imbued with the stern utilitarian spirit; for a decaying tree is perhaps more interesting than one in full vigour of growth. The starlings make their nests in

G

the upper knot-holes; or, lower down, the owl feeds her
young; and if you chance to pass near, and are not aware
of the ways of owls, you may fancy that a legion of
serpents are in the bushes, so loud and threatening is the
hissing noise made by the brood.   The woodpecker
comes for the insects that flourish on the dying giant; so
does the curious little tree-climber, running up the trunk
like a mouse; and in winter, when insect-life is scarce, it
is amusing to watch there the busy tomtit.   He hangs
underneath a dead branch, head downwards, as if walking
on a ceiling, and with his tiny but strong bill chips off a
fragment of the loose dead bark.   Under this bark, as he
well knows, woodlice and all kinds of creeping things
make their home.   With the fragment he flies to an
adjacent twig, small enough to be grasped by his claws
and so give him a firm foothold.   There he pecks his
morsel into minute pieces and lunches on the living con-
tents.   Then, with a saucy chuckle of delight in his own
cleverness, he returns to the larger bough for a fresh
supply.   As the bough decays the bark loosens, and is
invaded by insects which when it was green could not
touch it.

For the acorns the old oak still yields come rooks,
pigeons, and stately pheasants, with their glossy feathers
shining in the autumn sun.   Thrushes carry wild hedge-
fruit up on the broad platform formed by the trunk where
the great limbs divide, and pecking it to pieces, leave the

seeds. These take root in the crevices which widen out underneath into a mass of soft decaying 'touchwood;' and so from the crown of the tree there presently stream downwards long trailing briers, bearing in June the sweet wild roses and in winter red oval fruit. Ivy comes creeping up, and in its thick warm coverts nests are built. Below, among the powdery 'touchwood' which lines the floor of this living hut, great fungi push their coloured heads up to the light. And here you may take shelter when the rain comes unexpectedly pattering on the leaves, and listen as it rises to a roar within the forest. Sometimes wild bees take up their residence in the hollow, slowly filling it with comb, buzzing busily to and fro; and then it is not to be approached so carelessly, though so ready are all creatures to acknowledge kindness that ere now I have even made friends with the inhabitants of a wasp's nest.

A thick carpet of dark green moss grows upon one side of the tree, and over it the tall brake fern rears its yellow stem. In the evening the goat-sucker or nightjar comes with a whirling phantom-like flight, wheeling round and round: a strange bird, which will roost all day on a rail, blinking or sleeping in the daylight, and seeming to prefer a rail or a branch without leaves to one that affords cover. Here also the smaller bats flit in the twilight, and, if you stand still, will pursue their prey close to your head, wheeling about it so that you may knock them down with

your hand if you wish.   The labouring people call the bat
'bat-mouse.'   Here also come many beetles ; and sometimes on a summer's day the swallows will rest from their
endless flight on the dying upper branches, for they too
like a bough clear or nearly clear of leaves.   All the year
through the hollow tree is haunted by every kind of living
creature, and therefore let us hope it may yet be permitted
to linger awhile safe from the axe.

The lesser roots of the elm are porous like cane, and
are sometimes smoked as cigars by the ploughboys.   The
leaf of the coltsfoot, which grows so luxuriantly in many
places and used to be regularly gathered and dried by the
lower classes for the pipe, is now rarely used since the
commoner tobaccos have become universally accessible.

Often and often, when standing in a meadow gateway
partly hidden by the bushes, watching the woodpecker
on the ant-hills, of whose eggs, too, the partridges are so
fond (so that a good ant year, in which their nests are
prolific, is also a good partridge year), you may, if you are
still, hear a slight faint rustle in the hedge, and by-and-bye
a weasel will steal out.   Seeing you he instantly pauses,
elevates his head, and steadily gazes : move but your eyes
and he is back in the hedge ; remain quiet, still looking
straight before you as if you saw nothing, and he will
presently recover confidence, and actually cross the gateway almost under you.

This is the secret of observation : stillness, silence,

and apparent indifference. In some instinctive way these
wild creatures learn to distinguish when one is or is not
intent upon them in a spirit of enmity ; and if very near,
it is always the eye they watch. So long as you observe
them, as it were, from the corner of the eyeball, sideways,
or look over their heads at something beyond, it is well.
Turn your glance full upon them to get a better view,
and they are gone.

When waiting in a dry ditch with a gun on a warm
autumn afternoon for a rabbit to come out, sometimes a
bunny will suddenly appear at the mouth of a hole which
your knee nearly touches. He stops dead, as if petrified
with astonishment, sitting on his haunches. His full dark
eye is on you with a gaze of intense curiosity ; his nostrils
work as if sniffing ; his whiskers move ; and every now
and then he thumps with his hind legs upon the earth
with a low dull thud. This is evidently a sign of great
alarm, at the noise of which any other rabbit within
hearing instantly disappears in the 'bury.' Yet there
your friend sits and watches you as if spell-bound, so long
as you have the patience neither to move hand or foot,
nor to turn your eye. Keep your glance on a frond of
the fern just beyond him, and he will stay. The instant
your eye meets his or a finger stirs, he plunges out of
sight.

It is so also with birds. Walk across a meadow
swinging a stick, even humming, and the rooks calmly

continue their search for grubs within thirty yards ; stop
to look at them, and they rise on the wing directly.   So,
too, the finches in the trees by the roadside.   Let the
wayfarer pass beneath the bough on which they are sing-
ing, and they will sing on, if he moves without apparent
interest ; should he pause to listen, their wings glisten in
the sun as they fly.

The meadows lead down to the shores of the mere,
and the nearest fields melt almost insensibly into the green
margin of the water, for at the edge it is so full of flags,
and rushes, and weeds, as at a distance to be barely dis-
tinguishable there from the sward.   As we approach, the
cuckoo sings passing over head ; 'she cries as she flies' is
the common country saying.

I used to imagine that the cuckoo was fond of an
echo, having noticed that a particular clump of trees overhanging some water, the opposite bank of which sent
back a clear reply, was a specially favourite resort of that
bird.   The reduplication of the liquid notes, as they
travelled to and fro, was peculiarly pleasant : the water,
perhaps, lending, like a sounding-board, a fulness and
roundness to her song.   She might possibly have fancied
that another bird was answering ; certainly she 'cried'
much longer there than in other places.   Morning after
morning, and about the same time—eleven o'clock—a
cuckoo sang in that group of trees, from noting which I
was led to think that perhaps the cuckoo, though

apparently wandering aimlessly about, really has more method and regularity in her habits than would seem.

Country people will have it that cuckoos are growing scarcer every year, and do not come in the numbers they formerly did ; and, whether it be the chance of unfavourable seasons or other causes, it is certainly the fact in some localities. I recollect seeing as many as four at once in a tall elm—a tree they love—all crying and gurgling, as it were, in the throat together ; this was some years since, and that district is now much less frequented by these birds.

There was a superstition that where or in whatever condition you happened to be when you heard the cuckoo the first time in the spring, so you would remain for the next twelvemonth ; for which reason it was a misfortune to hear her first in bed, since it might mean a long illness. This, by-the-bye, may have been a pleasant fable invented to get milkmaids up early of a morning.

The number of coarse fish in the brook which flows out of the shallow mere bounding one edge of the keeper's domain of woods has, he thinks, very much decreased of recent years. When he first came here the stream seemed full of fish, notwithstanding very little care had till then been taken with their preservation. They used to net it once now and then, and he has seen a full hundredweight of fair-sized jack, perch, tench, etc., taken out of the water in a very short time, besides quantities of smaller fry which were put back again. But although the brook, so far as his jurisdiction goes, has since been comparatively well preserved, yet he feels certain the fish have diminished.

There are no chemical works to account for this with the subtle poison of their waste, neither are there mills to prevent the fish coming up—perhaps it would be better if there were some mills, as they would stop the fish going down. I have noticed that where old water-wheels have ceased working the fish have almost disappeared. This, of course, may be but a purely local phenomenon, but it is certainly the case in some districts. Comparatively little wheat now is ground in rural places; the greater portion is carried away to the towns and turned into flour by steam. So that in walking up a brook you will now and then come upon an ancient mill whose business has departed: the fabric itself is tenanted by two or three cottage families, and their garden covers the site of the old mill-pond. In the depths of that pool there were formerly plenty of fish, with deep dark spots in which to hide. Their natural increase was not swept away by floods; neither could they wander, because of the dam and grating. They were also under the eye of the miller, and so preserved. But when the dam was levelled and the stream allowed to follow its course, this resting-place, so to say, was abolished, and the fish dispersed were lost or captured.

Upon the particular brook which I have now in view

there are no mills ; but there used to be several large
ponds—distinct from the stream, yet communicating by
a narrow channel. These likewise sheltered the fish, and
were favourable to their propagation. Improvements,
however, have swept them away ; they are filled up, every
inch of ground having become valuable for agricultural
purposes. Then there were vast ditches running up
beside the hedgerows, and ending in the brook ; perfect
storehouses these of all aquatic life. Fish used to go up
them for shelter (they were as deep or deeper than the
brook itself, and it was a good jump for a man across),
and to feed on the insects blown off the overhanging
trees and bushes, or brought down by the streamlet draining the field above. Wild duck made their nests among
the rushes, sitting there while their beautiful consorts, the
mallards, swam lonely in the mere. Moorhens were busy
in the weeds, or came out to feed upon the sward.

Such great ditches are now filled up, and drains take
their place. It is better so, no doubt, in a purely utilitarian sense, but the fish haunt the spot no more. Some
of the reaches of the brook, where the ground was flat
and boggy, used to resemble a long narrow lake, extremely
shallow, with the deeper current running yards away from
the shore : and here the snipe came in the winter. But
the banks are now made up higher by artificial means,
and the marsh is dry. All these changes diminish those
aquatic nooks and corners in which fish love to linger.

WILD DUCK AND MOORHEN.

Finally came the weed said to have been imported from America, pushing its way up stream, and filling it with an abominable mass of vegetable matter that no fish could enter.   Hereabouts, however, this pest has of late shown signs of exhaustion—it does not grow with its former vigour, and its progress seems checked. The brook, after winding for several miles, the lower course being beyond the keeper's boundaries, empties itself into a canal ; before the canal was made it ran much farther, and itself increased in volume almost to a river. Now this canal is fished day and night by people on the tow-path : there is nominally a close-time, but no one observes it, and the riparian owners, having discovered that they had a right so to do, net it mercilessly. The consequence is that the fish which go down the stream and enter the canal are speedily destroyed, while the canal on its part sends no fish to the upper waters. This is how the decrease of fish is accounted for, and it is the same with perhaps half a dozen other brooks in the same locality, all of which now fall into the canal, which is so incessantly plied with rod and net and nightline that little escapes.

CHAPTER V.

Some of his Subjects: Dogs, Rabbits, 'Mice, and such small Deer.'

WHEN a dog, young and yet unskilled, follows his master across the meadows, it often happens that he meets with difficulties which sorely try the capacity of the inexperienced brain. The two came to a broad deep brook. The man glances at the opposite bank, and compares in his mind the distance to the other side with other distances he has previously leaped. The result is not quite satisfactory; somehow a latent doubt develops itself into a question of his ability to spring over. He cranes his neck, looks at the jump sideways to get an angular measurement, retires a few paces to run, shakes his head, deliberates, instinctively glances round as if for assistance or advice, and presently again advances to the edge. No; it will not do. He recalls to mind the division of space into yards, feet, and inches, and endeavours to apply it without a rule to the smooth surface of the water. He can judge a yard on the grass, because there is something to fix the eye on —the tall bennet or the buttercup yonder; but the water affords no data.

On second thoughts, yes—even the smooth flowing

current has its marks.   Here, not far from the steep bank
is a flag, bowed or broken, whose pennant-like tongue of
green floats just beneath the surface, slowly vibrating to
and fro, as you wave your hand in token of farewell.
This is mark one—say three feet from the shore.

Somewhat farther there is a curl upon the water, not
constant, but coming every few seconds in obedience to
the increase or decrease of the volume of the stream,
which there meets with some slight obstacle out of sight.
For, although the water appears level and unvarying, it
really rises and sinks in ever so minute a degree with a
rhythmic alternation.   If you will lie down on the sward,
you may sometimes see it by fixing a steady gaze upon
the small circular cave where the gallery of a water-rat
opens on this the Grand Canal of his Venice.   Into it
there rises now and again a gentle swell—barely perceptible—a faint pulse rising and falling.   The stream is
slightly fuller and stronger at one moment than another ;
and with each swell the curl, or tiny whirlpool, rotates
above the hidden irregularity of the bottom.   If you sit by
the dam higher up the brook, and watch the arch of the
cataract rolling over, it is perhaps more visible.   Every
now and then a check seems to stay the current momentarily : and at night, when it is perfectly still, listening to
the murmur of the falling water from a distance, under
the apple-trees in the garden it runs a scale—now up,
then down ;. each variation of volume changing the

musical note. This faint undulation is more visible in some brooks than others.

A third mark is where a branch, as it was carried along, grounded on a shallow spot; and one mast, as it were, of the wreck protruding upwards, catches the stray weeds as they swim down and holds them. Thus, step by step, the mind of the man measures the distance, and assures him that it is a little beyond what he has hitherto attempted; yet will not extra exertion clear it?—for having once approached the brink, shame and the dislike of giving up pull him forward. He walks hastily twenty yards up the brook, then as many the other way, but discovers no more favourable spot; hesitates again; next carefully examines the tripping place, lest the turf, undermined, yield to the sudden pressure, as also the landing, for fear of falling back. Finally he retires a few yards, and pauses a second and runs. Even after the start, uncertain in mind and but half resolved, it is his own motion which impels the will, and he arrives on the opposite shore with a sense of surprise. Now comes the dog, and note his actions; contrast the two, and say which is instinct, which is mind.

The dog races to the bank—he has been hitherto hunting in a hedge and suddenly misses his master—and, like his lord, stops short on the brink. He has had but little experience in jumping as yet; water is not his natural element, and he pauses doubtfully. He looks across earnestly, sniffs the air as if to smell the distance, then whines in distress of mind. Presently he makes a movement to spring, checks it, and turns round as if looking for advice or encouragement. Next he runs back a

DOG AT STREAM.

short way as if about to give it up; returns, and cranes over the brink; after which he follows the bank up and down, barking in excitement, but always coming back to the original spot. The lines of his face, the straining eye, the voice that seems struggling to articulate in the throat, the attitude of the body,—all convey the idea of intense

desire which fear prevents him from translating into action.
There is indecision—uncertainty—in the nervous grasp
of the paws on the grass, in the quick short coursings to
and fro.   Would infallible instinct hesitate?   He has no
knowledge of yards, feet, and inches—yet he is clearly trying to judge the distance.   Finally, just as his master
disappears through a gateway, the agony of his 'mind'
rises to the highest pitch.   He advances to the very brink
—he half springs, stays himself, his hinder paws slip down
the steep bank, he partly loses his balance, and then
makes a great leap, lights with a splash in mid-stream,
and swims the remainder with case.   There is, at least, a
singular coincidence in the outward actions of the two.

The gamekeeper, with dogs around him from morning
till night, associated with them from childhood, has no
doubts upon the matter whatever, but with characteristic
decision is perfectly certain that they think and reason in
the same way as human beings, though of course in a
limited degree.   Most of his class believe, likewise, in the
reasoning power of the dog: so do shepherds ; and so,
too, the labourers who wait on and feed cattle are fully
persuaded of their intelligence, which, however, in no way
prevents them throwing the milking-stool at their heads
when unruly.   But the concession of reason is no guarantee
against ill-usage, else the labourer's wife would escape.

The keeper, without thinking it perhaps, affords a
strong illustration of his own firm faith in the mind of the

dog. His are taught their proper business thoroughly ; but there it ends. ' I never makes them learn no tricks,' says he, ' because I don't like to see 'em made fools of.' I have observed that almost all those whose labour lies in the field, and who go down to their business in the green meadows, admit the animal world to a share in the faculty of reason. It is the cabinet thinkers who construct a universe of automatons.

No better illustration of the two modes of observation can be found than in the scene of Goethe's ' Faust ' where Faust and Wagner walking in the field are met by a strange dog. The first sees something more than a mere dog ; he feels the presence of an intelligence within the outward semblance—in this case an evil intelligence, it is true, but still a something beyond mere tail and paws and ears. To Wagner it is a dog and nothing more—that will sit at the feet of his master and fawn on him if spoken to, who can be taught to fetch and carry or bring a stick ; the end, however, proves different. So one mind sees the outside only ; another projects itself into the mind of the creature, be it dog, or horse, or bird.

Experience certainly educates the dog as it does the man. After long acquaintance and practice in the field we learn the habits and ways of game—to know where it will or not be found. A young dog in the same way dashes swiftly up a hedge, and misses the rabbit that, hearing him coming, doubles back behind a tree or stole ; an

old dog leaves nothing behind him, searching every corner.
This is acquired knowledge. Neither does all depend
upon hereditary predisposition as exhibited in the various
breeds—the setter, the pointer, the spaniel, or greyhound
—and their especial drift of brain ; their capacity is not
wholly confined to one sphere. They possess an initiating
power—what in man is called originality, invention, discovery : they make experiments.

I had a pointer that exhibited this faculty in a curious
manner. She was weakly when young, and for that
reason, together with other circumstances, was never
properly trained : a fact that may perhaps have prevented
her 'mind' from congealing into the stolidity of routine.
She became an outdoor pet, and followed at heel everywhere. One day some ponds were netted, and of the fish
taken a few chanced to be placed in a great stone trough
from which cattle drank in the yard—a common thing in
the country. Some time afterwards, the trough being
foul, the fish—they were roach, tench, perch, and one
small jack—were removed to a shallow tub while it was
being cleansed. In this tub, being scarcely a foot deep
though broad, the fish were of course distinctly visible,
and at once became an object of the most intense interest
to the pointer. She would not leave it ; but stood watching every motion of the fish, with her head now on one
side, now on the other. There she must have remained
some hours, and was found at last in the act of removing

POINTER AND FISH.

them one by one and laying them softly, quite unhurt, on the grass.

I put them back into the water, and waited to see the result. She took a good look, and then plunged her nose right under the surface and half-way up the neck, completely submerging the head, and in that position groped about on the bottom till a fish came in contact with her mouth and was instantly snatched out. Her head must have been under water each time nearly a minute, feeling for the fish. One by one she drew them out and placed

them on the ground, till only the jack remained. He
puzzled her, darting away swift as an arrow and seeming
to anticipate the enemy. But after a time he, too, was
captured.

They were not injured—not the mark of a tooth was
to be seen—and swam as freely as ever when restored
to the water. So soon as they were put in again the
pointer recommenced her fishing, and could hardly be got
away by force. The fish were purposely left in the tub.
The next day she returned to the amusement, and soon
became so dexterous as to pull a fish out almost the instant her nose went under water. The jack was always
the most difficult to catch, but she managed to conquer
him sooner or later. When returned to the trough, however, she was done—the water was too deep. Scarcely
anything could be imagined apparently more opposite to
the hereditary intelligence of a pointer than this; and
certainly no one attempted to teach her, neither did she
do it for food. It was an original motion of her own: to
what can it be compared but mind proceeding by experi-
ment? They can also adjust their conduct to circumstances, as when they take to hunting on their own
account: they then generally work in couples.

If a spaniel, for instance, one of those allowed to lie
loose about farmhouses, takes to hunting for herself, she is
almost always found to meet a canine friend at a little
distance from the homestead. It is said that spaniels

when they go off like this never bark when on the heels
of a rabbit, as they would do if a sportsman was with
them and the chase legitimate. This suppression of what
must be an almost uncontrollable inclination shows no
little intelligence. If they gave tongue, they would be
certainly detected, and as certainly thrashed. To watch
the sneaking way in which a spaniel will come home after
an unlawful expedition of this kind is most amusing.
She makes her appearance on the road or footpath so as
not to look as if coming from the hedges, and enters at
the back ; or if any movement be going on, as the driving
of cattle, she will join in it, displaying extraordinary zeal
in assisting : anything to throw off suspicion.

Of all sport, if a man desires to widen his chest, and
gain some idea of the chase as it was in ancient days, let
him take two good greyhounds and 'uncouple at the
timorous flying hare,' following himself on foot. A race
like this over the elastic turf of the downs, inhaling with
expanded lungs air which acts on the blood as strong
drink on the brain, stimulating the pulse, and strengthen-
ing every fibre of the frame, is equal to another year of
life. Coursing for the coursing's sake is capital sport.
A hare when sorely tried with the hot breath from the
hounds' nostrils on his flanks, will sometimes puzzle them
by dashing round and round a rick. Then in sweeping
circles the trio strain their limbs, but the hare, having at the
corners the inner side and less ground to cover, easily

keeps just ahead. This game lasts several minutes, till presently one of the hounds is sharp enough to dodge back and meet the hare the opposite way. Even then his quick eye and ready turn often give him another short breathing space by rushing away at a tangent.

Rabbits, although of 'low degree' in comparison with the pheasant, really form an important item in the list of the keeper's charges. Shooting generally commences with picking out the young rabbits about the middle or towards the end of the hay harvest, according as the season is early or late. Some are shot by the farmers, who have the right to use a gun, earlier than this, while they still disport in the mowing grass. It requires experience and skill to select the young rabbit just fit for table from the old bucks, the does which may yet bring forth another litter, and those little bunnies that do not exceed the size of rats.

The grass conceals the body of the animal, and nothing is visible beyond the tips of the ears; and at thirty yards distance one pair of ears is very like another pair. The developed ear is, however, less pointed than the other; and in the rabbit of a proper size they are or seem to be wider apart. The eye is also guided by the grass itself and the elevation of the rabbit's head above it when lifted in alarm at a chance sound: if the animal is full grown of course the head stands higher. In motion the difference is at once seen; the larger animal's back and flanks show boldly, while the lesser seems to slip through the grass. By these signs, and by a kind of instinct which grows upon one when always in the field, it is possible to distinguish between them even in tall grass and in the gloaming.

This sort of shooting, if it does not afford the excitement of the pheasant battue, or require the alertness necessary in partridge killing, is not without its special pleasures. These are chiefly to be attributed to the genial warmth of the weather at that season, when the reapers have only just begun to put the tall corn to the edge of their crooked swords, and one can linger by the hedge-side without dread of wintry chills.

The aftermath in which the rabbits feed is not so tall as the mowing grass, and more easy and pleasant to walk through, though it is almost devoid of flowers. Neither does it give so much shelter; and you must walk close to the hedge, gliding gently from bush to bush, the slower the better. Rabbits feed several times during the day— i.e. in the very early morning, next about eleven o'clock, again at three or four, and again at six or seven. Not that every rabbit comes out to nibble at those hours, but about that time some will be seen moving outside the buries.

As you stroll beside the hedge, brushing the boughs, a rabbit feeding two hundred yards away will lift his head inquiringly from the grass. Then stop, and remain still as the elm tree hard by. In a minute or two, reassured, the ears perked up so sharply fall back, and he feeds again. Another advance of ten or twenty yards, and up go the ears—you are still till they drop once more. The rabbit presently turns his back towards you, sniffing about for the tenderest blades; this is an opportunity, and an advance of forty or fifty paces perhaps is accomplished. Now, if you have a rook-rifle you are near enough; if a smooth-bore, the same system of stalking must be carried farther yet. If you are patient enough to wait when he takes alarm, and only to advance when he feeds, you are pretty sure to 'bag' him.

Sometimes, when thus gliding with stealthy tread, another rabbit will suddenly appear out of the ditch within easy reach; it is so quiet he never suspected the presence of an enemy. If you pause and keep quite still, which is the secret of all stalking, he will soon begin to feed, and the moment he turns his back towards you up goes the gun; not before, because if he sees your arm move he will be off to the ditch. True, a snap-shot might be made as he runs, which at first sight would appear more sportsmanlike than 'potting;' but it is not so, for it is ten chances to one that you do not kill him dead on the spot in the short distance he has to traverse. Perhaps the hind legs will be broken; well, then he will drag them along behind him, using the fore paws with astonishing rapidity and power. Before the second barrel can be emptied he will gain the shelter of the fern that grows on the edge of the bank and dive into a burrow, there to die in misery. So that it is much better to steadily 'pot' him. Besides which, if a rabbit dies in a burrow all the other animals in that particular burrow desert it till nature's scavengers have done their work. A dog cannot well be taken while stalking—not that dogs will not follow quietly, but because a rabbit, catching sight of a dog, is generally stricken with panic even if a hundred yards away, and bolts immediately.

I have seen a rabbit whose back was broken by shot drag itself ten yards to the ditch. If the forelegs are broken, then he is helpless : all the kicks of the hind legs only tumble him over and over without giving him much progress. The effects of shot are very strange, and sometimes almost inexplicable ; as when a hare which has received a pellet through the edge of the heart runs a quarter of a mile before dropping. It is noticed that hares and rabbits, hit in the vital organs about the heart, often run a considerable distance, and then, suddenly in the midst of their career, roll head over heels dead. Both hares and rabbits are occasionally killed with marks of old shot wounds, but not very often, and they are but of a slight character—the pellets are found just under the skin, with a kind of lump round them. Shot holes through the ears are frequently seen, of course doing no serious harm.

Now and then a rabbit hit in the head will run round

and round in circles, making not the slightest attempt to
escape.   The first time I saw this, not understanding it, I
gave the creature the second barrel ; but next time I let
the rabbit do as he would.   He circled round and round,
going at a rapid pace.   I stood in his way, and he passed
between my legs.   After half a dozen circles the pace grew
slower.   Finally, he stopped, sat up quite still for a minute
or so, and then drooped and died.   The pellet had struck
some portion of the brain.

I once, while looking for snipe with charges of small
shot in the barrels, roused a fine hare, and fired without
apparent effect.   But after crossing about half of the field
with a spaniel tearing behind, he began to slacken speed,
and I immediately followed.   The hare dodged the spaniel
admirably, and it was with the utmost difficulty I secured
him (refraining from firing the second barrel on purpose).
He had been stopped by one single little pellet in the
great sinew of the hind leg, which had partly cut it through.
Had it been a rabbit he would certainly have escaped into
a bury, and there, perhaps died, as shot wounds frequently
fester : so that in stalking rabbits, or waiting for them
behind a tree or bush, it is much better to take a steady
aim at the head, and so avoid torturing the creature.

' Potting ' is hardly sport, yet it has an advantage to
those who take a pleasure in observing the ways of bird
and animal.   There is just sufficient interest to induce one
to remain quiet and still, which is the prime condition of

seeing anything ; and in my own case the rabbits so
patiently stalked have at last often gone free, either from
their own amusing antics, or because the noise of the explosion would disturb something else under observation.
In winter it is too cold ; then you step quietly and yet
briskly up to a fence or a gateway, and glance over, and
shoot at once ; or with the spaniels hunt the bunnies from
the fern upon the banks, yourself one side of the hedge
and the keeper the other.

In excavating his dwelling, the rabbit, thoughtless of
science, constructs what may be called a natural auditorium
singularly adapted for gathering the expiring vibrations of
distant sound.   His round tunnel bored in a sandy bank is
largest at the opening, like the mouth of a trumpet, and
contracts within—a form which focusses the undulations of
the air.   To obtain the full effect the ear should listen
some short way within ; but the sound, as it is thrown
backwards after entering, is often sufficiently marked to be
perceptible when you listen outside.   The great deep
ditches are dry in summer ; and though shooting be not
the object, yet a gun for knocking over casual vermin is a
pleasant excuse for idling in a reclining position shoulder-
high in fern, hidden like a skirmisher in such an entrench-
ment.   A mighty root bulging from the slope of the bank
forms a natural seat.   There is a cushion of dark green
moss to lean against, and the sand worked out from the
burrows—one nearly on a level with the head and another

lower down—has here filled up the ditch to some height, making a footstool.

In the ditch lie numbers of last year's oak leaves, which so sturdily resist decay. All the winter and spring they were soaked by the water from the 'land-springs'—as those which only run in wet weather are called—draining into it, and to that water they communicated a peculiar flavour, slightly astringent. Even moderate-sized stream-lets become tainted in the latter part of the autumn by the mass of leaves they carry down, or filter through, in wood-land districts. Often the cottagers draw their water from a small pool filled by such a ditch, and coated at the bottom with a thick layer of decomposing leaves. The taste of this water is strong enough to overcome the flavour of their weak tea, yet they would rather use such water than walk fifty yards to a brook. It must, however, be admitted that the brooks at that time are also tinctured with leaf, and there seems to be no harm in it.

Out from among these dead leaves in the ditch pro-trudes a crooked branch fallen long since from the oak, and covered with grey lichen. On the right hand a tangled thicket of bramble with its uneven-shaped stems closes the spot in, and on the left a stole of hazel rises with the parasitical 'hardy fern' fringing it near the earth. The outer bark of the hazel is very thin; it is of a dark mottled hue; bruise it roughly, and the inner bark shows a bright green. The lowly ivy creeps over the bank—its

leaves with five angles, and variegated with grey streaks.
Through the hawthorn bushes above comes a faint but
regular sound—it is the parting fibres of the grass-blades
in the meadow on the other side as the cows tear them
apart, steadily eating their way onwards.   The odour of
their breath floats heavy on the air.   The sun is sinking,
and there is a hush and silence.

   But the rabbit-burrow here at my elbow is not silent ;
it seems to catch and heighten faint noises from a distance.
A man is walking slowly home from his work up the lane
yonder ; the fall of his footsteps is distinctly rendered by
the hole here.   The dull thuds of a far-off mallet or 'bitel'
(beetle) driving in a stake are plainly audible.   The thump-
thump of a horse's hoofs cantering on the sward by the
roadside, though deadened by the turf, are reproduced or
sharpened.   Most distinct of all comes the regular sound
of oars against the tholepins or rowlocks of a boat moving
on the lake many fields away.   So that in all probability
to the rabbit his hole must be a perfect ' Ear of Dionysius,'
magnifying a whisper—unless, indeed, its turns and wind-
ings confuse the undulations of sound.   It is observable
that before the rabbit ventures forth he stays and listens
just within the entrance of his burrow, where he cannot *see*
any danger unless absolutely straight before him—a habit
that may have unconsciously grown up from the apparent
resonance of sound there.

   Sitting thus silently on the root of the oak, presently

I hear a slight rustling among the dead leaves at the bottom of the ditch. They heave up as if something was pushing underneath ; and after a while, as he comes to the heap of sand thrown out by the rabbits, a mole emerges, and instantly with a shiver, as it were, of his skin throws off the particles of dust upon his fur as a dog fresh from the water sends a shower from his coat. The summer weather having dried the clay under the meadow turf and

TRAPS.

made it difficult to work, he has descended into the ditch, beneath which there is still a certain moistness, and where he can easily bore a tunnel.

It is rather rare to see a mole above ground ; fortunately for him he is of diminutive size, or so glossy a fur would prove his ruin. As it is, every other old pollard willow tree along the hedge is hung with miserable moles, caught in traps, and after death suspended — like criminals swinging on a gibbet —

from the end of slender willow boughs. Moles seem to

breed in the woods : first perhaps because they are less disturbed there, next because under the trees the earth is usually softer, retains its moisture longer, and is easier to work. From the woods their tracks branch out, ramifying like the roads which lead from a city. They have in addition main arteries of traffic, king's highways, along which they will journey one after the other ; so that the mole-catcher, if he can discover such a road, slaughters many in succession. The heaps they throw up are awkward in mowing grass, the scythe striking against them ; and in consequence of complaints of their rapid multiplication in the woods the keeper has to employ men to reduce their numbers. It is curious to note how speedily the mole buries himself in the soil ; it is as if he suddenly dived into the earth.

Another slight rustling—a pause, and it is repeated ; this time on the bank, among the dry grass. It is mice ; they have a nervous habit of progressing in sharp, short stages. They rush forward seven or eight inches with lightning-like celerity—a dun streak seems to pass before your eye ; then they stop short a moment or two, and again make another dash. This renders it difficult to observe them, especially as a single dead brown leaf is sufficient to hide one. It is so silent that they grow bold, and play their antics freely, darting to and fro, round and under the stoles, chasing each other. Sometimes they climb the bushes, running along the upper surface of the boughs that chance to be nearly horizontal. Once on a hawthorn branch in a hedge I saw a mouse descending with an acorn; he was, perhaps, five feet from the ground, and how and from whence he had got his burden was rather puzzling at first. Probably the acorn, dropping from the tree, had been caught and held in the interlacing of the bush till observed by the keen, if tiny, eyes below.

Mice have a magical way of getting into strange places. In some farmhouses they still use the ancient, old-fashioned lanterns made of tin—huge machines intended for a tallow candle, and with plates of thin translucent horn instead of glass. They are not wholly despicable; since if set on the ground and kicked over by a recalcitrant cow in the sheds, the horn does not break as glass would. These lanterns, having a handle at the top, are by it hung up to the beam in the kitchen; and sometimes to the astonishment of the servants in the quiet of the evening, they are found to be animated by some motive power, swinging to and fro and partly turning round. A mouse has got in—for the grease; but how? that is the 'wonderment,' as the rustic philosophers express it; for, being hung from the beam, eight or nine feet from the stone-flagged floor, there seems no way of approach for the mouse except by 'walking' on the ceiling or along and partly underneath the beam itself. If so, it would seem to be mainly by the propulsive power exerted previous to starting on the trip—just as a man can get a little way up the perpendicular side of a rick by

running at it. Occasionally, no doubt, the mouse has
entered when the lantern has been left opened while
lighted on the ground, and so got shut in; but mice have
been found in lanterns cobwebbed from long disuse.

Suddenly there peeps out from the lower rabbit-hole
the stealthy reddish body of a weasel. I instinctively
reach for the gun leaning against the bank, and immediately the spell is broken. The mice rush to their holes,
the weasel darts back into the bowels of the earth, a
rabbit that has quietly slipped out unseen into the grass
bounds with eager haste to cover, and out of the oak overhead there rises, with a great clatter of wings, a wood-
pigeon that had settled there.

When the pale winter sunshine falls upon the bare
branches of an avenue of elms—such as so often ornament
parks—they appear lit up with a faint rosy colour, which
instantly vanishes on the approach of a shadow. This
shimmering mirage in the boughs seems due to the
myriads of lesser twigs, which at the extremities have a
tinge of red, invisible at a distance till the sunbeams
illuminate the trees. Beyond this passing gleam of colour,
nothing relieves the blackness of the January landscape,
except here and there the bright silvery bark of the birch.

For several seasons now in succession the thrush has
sung on the shortest days, as though it were spring; a
little later, in the early mornings, the blackbird joins,
filling the copse with a chorus at the dawn. But, if the

I

wind turns to east or north, the rooks perch on the oaks
in the hedgerows in the middle of the day, puffing out
their feathers and seeming to abandon all search for food,
as if seized with uncontrollable melancholy.   Hardy as
these birds are, a long frost kills them in numbers,
principally by slow starvation.   They die during the
night, dropping suddenly from their roosting-place on the
highest boughs of the great beech-trees, with a thud
distinctly heard in the silence of the woods.   The leaves
of the beech decay so gradually as to lie in heaps beneath
for months, filling up the hollows, so that an unwary
passer-by may plunge knee-deep in leaves.   Rooks when
feeding usually cross the field facing the wind, perhaps to
prevent the ruffling of their feathers.

Wood-pigeons have apparently much increased in
numbers of recent years ; they frequent sheltered spots
where the bushes diminish the severity of the frost.
Sometimes on the hills at a lonely farmhouse, where the
bailiff has a long-barrelled ancient fowling-piece, he will
lay a train of grain for them, and with a double charge of
shot, kill many at a time.

Men have boasted of shooting twenty at once.   But
with an ordinary gun it is not credible ; and the statement,
without wilful exaggeration, may arise from confusion in
counting, for it is a fact that some of the older uneducated
country labourers cannot reckon correctly.   It is not unusual in parishes to hear of a cottage woman who has had

twenty children. Upon investigation the real number is found to be sixteen or seventeen, yet nothing on earth will convince the mother that she has not given birth to a score. They get hazy in figures when exceeding a dozen.

SHOOTING WOOD-PIGEONS BY MOONLIGHT.

A pigeon is not easily brought down—the quills are so stiff and strong that the shot, if it comes aslant, will glance off. Many pigeons roost in the oaks of the hedges, choosing by preference one well hung with ivy, and when it is a moonlit night afford tolerable sport. It requires a

gun on each side of the hedge. A stick flung up awakes
the birds; they rise with a rush and clatter, and in the
wildness of their flight and the dim light are difficult to
hit. There is a belief that pigeons are partially deaf. If
stalked in the daytime they take little heed of footsteps or
slight noises which would alarm other creatures; but, on
the other hand, they are quick of eye, and are gone directly
anything suspicious appears in sight. You may get quite
under them and shoot them on the bough at night. It is
not their greater wakefulness but the noise they make in
rising which renders them good protectors of preserves; it
alarms other birds and can be heard at some distance.

When a great mound and hedgerow is grubbed up,
the men engaged in the work often anticipate making a
considerable bag of the rabbits, whose holes riddle it in
every direction, thinking to dig them out even of those
innermost chambers whence the ferret has sometimes been
unable to dislodge them. But this hope is almost always
disappointed; and when the grub-axe and spade have
laid bare the 'buries' only recently teeming with life, not
a rabbit is found. By some instinct they have discovered
the approach of destruction, and as soon as the first few
yards of the hedge are levelled secretly depart. After a
'bury' has been ferreted it is some time before another
colony takes possession: this is seemingly from the intense
antipathy of the rabbit to the smell of the ferret. Even
when shot at and pressed by dogs, a rabbit in his hasty

rush will often pass a hole which would have afforded instant shelter because it has been recently ferreted.

At this season the labourers are busy with 'beetle' (pronounced 'bitel')—a huge mallet—and iron wedges, splitting the tough elm-butts and logs for firewood. In old times a cottager here and there with a taste for astrology used to construct an almanack by rule of thumb, predicting the weather for the ensuing twelve months from the first twelve days of January. As the wind blew on those days, so the prevailing weather of the months might be foretold. The aged men, however, say that in this divination the old style must be adhered to, for the sequence of signs and omens still follows the ancient reckoning, which ought never to have been interfered with.

CHAPTER VI.

His Enemies—Birds and Beasts of Prey—Trespassers.

THERE are other enemies of game life besides human poachers whose numbers must be kept within bounds to ensure successful sport.   The thirst of the weasel for blood is insatiable, and it is curious to watch the persistency with which he will hunt down the particular rabbit he has singled out for destruction.   Through the winding subterranean galleries of the 'buries' with their cross-passages, 'blind' holes and 'pop' holes (*i.e.* those which end in undisturbed soil, and those which are simply bored from one side of the bank to the other, being only used for temporary concealment), never once in the dark close caverns losing sight or scent of his victim, he pursues it with a species of eager patience.   It is generally a long chase.   The rabbit makes a dash ahead and a double or two, and then halts, usually at the mouth of a hole : perhaps to breathe.   By-and-by the weasel, baffled for a few minutes, comes up behind.   Instantly the rabbit slips over the bank outside and down the ditch for a dozen yards, and there enters the 'bury' again.   The weasel follows, gliding up the bank with a motion not unlike that of the snake ; for his body and neck are long and slender and his legs short.   Apparently he is not in haste, but rather lingers over the scent.   This is repeated five or six times, till the whole length of the hedgerow has been traversed— sometimes up and down again.   The chase may be easily observed by any one who will keep a little in the background.   Although the bank be tenanted by fifty other rabbits, past whose hiding-place the weasel must go, yet they scarcely take any notice.   One or two whom he has approached too closely bolt out and in again ; but as a mass the furry population remain quiet, as if perfectly aware that they are not yet marked out for slaughter.

At last, having exhausted the resources of the bank, the rabbit rushes across the field to a hedgerow, perhaps a hundred yards away.   Here the wretched creature seems to find a difficulty in obtaining admittance.   Hardly has he disappeared in a hole before he comes out again, as if the inhabitants of the place refused to give him shelter. For many animals have a strong tribal feeling, and their sympathy, like that of man in a savage state, is confined within their special settlement.

With birds it is the same : rooks, for instance, will not allow a strange pair to build in their trees, but drive them off with relentless beak, tearing down the half-formed nest, and taking the materials to their own use.   The sentiment, ' If Jacob take a wife of the daughters of Heth, what good shall my life do me ?' appears to animate the breasts of gregarious creatures of this kind. Rooks intermarry generation after generation ; and if a black lover brings home a foreign bride they are forced to build in a tree at some distance. Near large rookeries several such outlying colonies may be seen.

The rabbit, failing to find a cover, hides in the grass and dry rushes ; but across the meadow, stealing along the furrow, comes the weasel ; and, shift his place how he may, in the end, worn out and weary, bunny succumbs, and the sharp teeth meet in the neck behind the ear, severing the vein. Often in the end the rabbit runs to earth in a hole which is a *cul-de-sac*, with his back towards the pursuer. The weasel, unable to get at the poll, which is his desire, will mangle the hinder parts in a terrible manner—as will the civilised ferret under similar conditions. Now and then the rabbit, scratching and struggling, fills the hole in the rear with earth, and so at the last moment chokes off his assailant and finds safety almost in the death-agony. In the woods, once the rabbit is away from the ' buries,' the chase really does resemble a hunt ; from furze-bush to bracken, from fern to rough grass, round and round, backwards, doubling, to and fro, and all in vain.

At such times, eager for blood, the weasel will run right across your path, almost close enough to be kicked. Pursue him in turn, and if there be no hedge or hole near, if you have him in the open, he will dart hither and thither right between your legs, uttering a sharp short note of

A WEASEL HUNTING.

anger and alarm, something composed of a tiny bark and
a scream.   He is easily killed with a stick when you catch
him in the open, for he is by no means swift; but if a
hedge be near it is impossible to secure him.

Weasels frequently hunt in couples, and sometimes
more than two will work together.   I once saw five, and
have heard of eight.   The five I saw were working a sandy

bank drilled with holes, from which the rabbits in wild alarm were darting in all directions. The weasels raced from hole to hole and along the sides of the bank exactly like a pack of hounds, and seemed intensely excited. Their manner of hunting resembles the motions of ants; these insects run a little way very swiftly, then stop, turn to the right or left, make a short detour, and afterwards on again in a straight line. So the pack of weasels darted forward, stopped, went from side to side, and then on a yard or two, and repeated the process. To see their reddish heads thrust for a moment from the holes, then withdrawn to reappear at another, would have been amusing had it not been for the reflection that their frisky tricks would assuredly end in death. They ran their quarry out of the bank and into a wood, where I lost sight of them. The pack of eight was seen by a labourer returning down a woodland lane from work one afternoon. He told me he got into the ditch, half from curiosity to watch them, and half from fear—laughable as that may seem—for he had heard the old people tell stories of men in the days when the corn was kept for years in barns, and so bred hundreds of rats, being attacked by those vicious brutes. He said they made a noise, crying to each other, short sharp snappy sounds; but the pack of five I myself saw hunted in silence.

Stoats, though not so numerous as weasels, probably do quite as much injury, being larger, swifter, stronger, and very bold, sometimes entering sheds close to dwelling-houses. The labouring people—at least the elder folk—declare that they have been known to suck the blood of infants left asleep in the cradle upon the floor, biting the child behind the ear. They hunt in couples also—seldom in larger numbers. I have seen three at work together, and with a single shot killed two out of the trio. In elegance of shape they surpass the weasel, and the colour is brighter. Their range of destruction seems only limited by their strength : they attack anything they can manage.

The keeper looks upon weasel and stoat as bitter foes, to be ruthlessly exterminated with shot and gin. He lays to their charge deadly crimes of murder, the death of rabbits, hares, birds, the theft and destruction of his young broods, even occasional abstraction of a chicken close to his very door, despite the dogs chained there. They are not easily shot, being quick to take shelter at the sight of a dog, and when hard hit with the pellets frequently escaping, though perhaps to die. Both weasel and stoat, and especially the latter, will snap viciously at the dog that overtakes them, even when sore wounded, always aiming to fix their teeth in his nose, and fighting savagely to the last gasp. The keeper slays a wonderful number in the course of a year, yet they seem as plentiful as ever. He traps perhaps more than he shoots.

It is not always safe to touch a stoat caught in a trap ;

he lies apparently dead, but lift him up, and instantly his teeth are in your hand, and it is said such wounds sometimes fester for months. Stoats are tough as leather: though severely nipped by the iron fangs of the gin, struck on the head with the butt of the gun, and seemingly quite lifeless, yet, if thrown on the grass and left, you will often find on returning to the place in a few hours' time that the animal is gone. Warned by experiences of this kind, the keeper never picks up a stoat till 'settled' with a stick or shot, and never leaves him till he is nailed to the shed. Stoats sometimes emit a disgusting odour when caught in a trap. The keeper has no mercy for such vermin, though he thinks some of his feathered enemies are even more destructive.

Twice a year the hawks and other birds of prey find a great feast spread before them ; first, in the spring and early summer, when the hedges and fields are full of young creatures scarcely able to use their wings, and again in the severe weather of winter when cold and hunger have enfeebled them.

It is difficult to understand upon what principle the hawk selects his prey. He will pass by with apparent disdain birds that are within easy reach. Sometimes a whole cloud of birds will surround and chase him out of a field ; and he pursues the even tenour of his way unmoved, though sparrow and finch almost brush against his talons. Perhaps he has the palate of an epicure, and likes to vary

the dish of flesh torn alive from the breast of partridge,
chicken, or mouse.   He does not eat all he kills; he will

A HAWK PURSUED BY FINCHES AND SWALLOWS.

sometimes carry a bird a considerable distance and then
drop the poor thing.   Only recently I saw a hawk, pursued by twenty or thirty finches and other birds across a

ploughed field, suddenly drop a bird from his claws as he passed over a hedge. The bird fell almost perpendicularly, with a slight fluttering of the wings, just sufficient to preserve it from turning head-over-heels, and on reaching the hedge could not hold to the first branches, but brought up on one near the ground. It was a sparrow, and was not apparently hurt—simply breathless from fright.

All kinds of birds are sometimes seen with the tail feathers gone : have they barely escaped in this condition from the clutches of the hawk ? Blackbirds, thrushes, and pigeons are frequently struck : the hawk seems to lay them on the back, for if he is disturbed that is the position his victim usually remains in. Though hawks do not devour every morsel, yet as a rule nothing is found but the feathers—usually scattered in a circle. Even the bones disappear : probably ground vermin make away with the fragments.

The hawk is not always successful in disabling his prey. I have seen a partridge dashed to the ground, get up again, and escape. The bird was flying close to the ground when struck ; the hawk alighted on the grass a few yards farther in a confused way as if overbalanced, and before he could reach the partridge the latter was up and found shelter in a thick hedge.

The power to hover or remain suspended in one place in the air does not, as some have supposed, depend upon the assistance of the wind, against which the hawk inclines the plane of his wings like an artificial kite. He can accomplish the feat when the air is quite still and no wind stirring. Nor is he the only bird capable of doing this, although the others possess the power in a much less degree. The common lark sometimes hovers for a few moments low down over the young green corn, as if considering upon what spot to alight. The flycatcher contrives to suspend itself momentarily, but it is by a rapid motion of the wings, and is done when the first snap at the insect has failed. It is the rook that hovers by the assistance of the wind as he rises with his broad, flat wings over a hedge and meets its full force, which counterpoises his onward impetus and sustains him stationary, sometimes compelling him to return with the current.

Hawks have a habit of perching on the tops of bare poles or dead trees, and are there frequently caught in the gin the keeper sets for them. The cuckoo, which so curiously resembles the hawk, has the same habit, and will perch on a solitary post in the middle of a field, or on those upright stones sometimes placed for the cattle to rub themselves against. Though 'wild as a hawk' is a proverbial phrase, yet hawks are bold enough to enter gardens, and even take their prey from the ivy which grows over the gable of the house. The destruction they work among the young partridges in early summer is very great. The keeper is always shooting them, yet they come just the same, or nearly ; for, if he exterminates them one season others arrive from a distance. He is particularly careful to look out for their nests, so as to kill both the old birds and to prevent their breeding. There is little difficulty in finding the nest (which is built in a high tree) when the young get to any size; their cry is unmistakable and audible at some distance.

Against sparrow-hawk and kestrel, and the rarer kinds that occasionally come down from the mountains of the north or the west—the magazines of these birds—the keeper wages ceaseless war.

So too with jay and magpie; he shoots them down whenever they cross his path, unless, as is sometimes the case, specially ordered to save the latter. For the magpie of recent years has become much less common. Though still often seen in some districts, there are other localities where this odd bird is nearly extinct. It does not seem to breed now, and you may ask to be shown a nest in vain. A magpie's nest in an orchard that I knew of was thought so great a curiosity that every now and then people came to see it from a distance. In other places the bird may be frequently met with, almost always with his partner; and so jays usually go in couples, even in winter.

The jay is a handsome bird, whose chatter enlivens the plantations, and whose bright plumage contrasts pleasantly with the dull green of the firs. A pair will work a hedge in a sportsmanlike manner, one on one side,

the second on the other ; while the tiny wren, which creeps
through the bushes as a mouse through the grass, cowers
in terror, or slips into a knot-hole till the danger is past.
When the husbandman has sown his field with the drill,
hardly has he left the gateway before a legion of small
birds pours out from the hedgerows and seeks for the
stray seeds. Then you may see the jay hop out among
them with an air of utter innocence, settling on the larger
lumps of clay for convenience of view, swelling out his
breast in pride of beauty, jerking his tail up and down, as
if to say, Admire me. With a sidelong hop and two flaps
of the wing, he half springs, half glides to another coign
of vantage. The small birds, sparrows, chaffinches, green-
finches—instantly scatter swiftly right and left, not rising,
but with a hasty run for a yard or so. They know well
his murderous intent, and yet are so busy they only put
themselves just out of reach, aware that, unlike the hawk,
he cannot strike at a distance. This game will continue
for a long time ; the jay all the while affecting an utter
indifference, yet ever on the alert till he spies his chance.
It is the young or weakly partridges and pheasants that
fall to the jay and magpie.

The keeper also destroys owls—on suspicion. Now
and then some one argues with the keeper, assuring him
that they do not touch game, but this he regards as pure
sentimentalism. 'Look at his beak,' is his steady reply.
'Tell me that that there bill weren't made to tear a bird's

K

breast to bits? Just see here—all crooked and pointed:
why, an owl have got a hooked bill like an eagle. It
stands to reason as he must be in mischief.' So the poor
owls are shot and trapped, and nailed to the side of the
shed.

But upon the crow the full vials of the keeper's wrath
are poured, and not without reason. The crow among
birds is like the local professional among human poachers:
he haunts the place and clears everything—it would be
hard to say what comes amiss to him. He is the imper-
sonation of murder. His long, stout, pointed beak is a
weapon of deadly power, wielded with surprising force by
the sinewy neck. From a tiny callow fledgling, fallen out
of the thrush's nest, to the partridge or a toothsome young
rabbit, it is all one to him. Even the swift leveret is said
sometimes to fall a prey, being so buffeted by the sooty
wings of the assassin and so blinded by the sharp beak
striking at his eyes as to be presently overcome. For the
crow has a terrible penchant for the morsel afforded by
another's eyes: I have seen the skull of a miserable thrush,
from which a crow rose and slowly sailed away, literally
split as if by a chisel—doubtless by the blow that destroyed
its sight. Birds that are at all diseased or weakly—as
whole broods sometimes are in wet unkindly seasons—
rabbits touched by the dread parasite that causes the fatal
'rot,' the young pheasant straying from the coop, even the
chicken at the lone farmstead, where the bailiff only lives

and is in the fields all day—these are the victims of the crow.

Crows work almost always in pairs—it is remarkable that hawks, jays, magpies, crows, nearly all birds of prey, seem to remain in pairs the entire year—and when they have once tasted a member of a brood, be it pheasant, partridge, or chicken, they stay till they have cleared off the lot. Slow of flight and somewhat lazy of habit, they will perch for hours on a low tree, croaking and pruning their feathers ; they peer into every nook and corner of the woodlands, not like the swift hawk, who circles over and is gone and in a few minutes is a mile away. So that neither the mouse in the furrow nor the timid partridge cowering in the hedge can escape their leering eyes.

Therefore the keeper smites them hip and thigh whenever he finds them ; and if he comes across the nest, placed on the broad top of a pollard-tree—not in the branches, but on the trunk—sends his shot through it to smash the eggs. For if the young birds come to maturity they will remain in that immediate locality for months, working every hedge and copse and ditch with cruel pertinacity. In consequence of this unceasing destruction the crow has become much rarer of late, and its nest is hardly to be found in many woods. They breed in the scattered trees of the meadows and fields, especially where no regular game preservation is attempted, and where no

keeper goes his rounds.   Even to this day a lingering
superstition associates this bird with coming evil ; and I
have heard the women working in the fields remark that
such and such a farmer then lying ill would not recover,
for a crow had been seen to fly over his house but just
above the roof-tree.

Trespassers give him a good deal of trouble, for a
great wood seems to have an irresistible attraction for all
sorts of semi-Bohemians, besides those who come for
poaching purposes.   The keeper thinks it much more
difficult to watch a wood like this, which is continuous
and all in one, than it is to guard a number of detached
plantations, though in the aggregate they may cover an equal
area.   It is impossible to see into it any distance ; to
walk round it is a task of time.   A poacher may slink
from tree to tree and from thicket to thicket, and, unless
the dogs chance to sniff him out, may lie hidden in tangled
masses of fern and bramble, while the keeper passes not
ten yards away.   But plantations laid out in regular
order with broad open spaces, sometimes with small fields
between, do not afford anything like cover for human
beings.   If a man is concealed in one of these copses, and
finds that the keeper or his assistants are about to go
through it, he must move or be caught ; and in moving he
has to pass across an open space, and is nearly sure to be
detected.   In a continuous wood of large extent, if he
hears the keepers coming, he has but to slip as rapidly

THE POACHER.

and silently as possible to one side, and often has the pleasure to see them pass right over the spot where only recently he was lying.

Therefore, although a wood is much more beautiful from an artistic point of view, with its lovely greens in spring and yellow and browns in autumn, its shades and recesses and fern-strewn glades, yet if a gentleman desires to imitate the monarch who laid out the New Forest and plant wood, and his object be simply game, the keeper is of opinion that the somewhat stiff and trim plantations are preferable. They are generally of fir; and fir is the most difficult of trees to slip past, being decidedly of an obstructive turn. The boughs grow so close to the ground that unless you crawl you cannot go under them. The trunks—unlike those of many other trees — will flourish so near together that the extremities of the branches touch and almost interweave, and they are rough and unpleasant to push through. To shoot or trap, or use a net or other poacher's implement, is very difficult in a young fir plantation, because of this thickness of growth; so that in a measure the tree itself protects the game. Then the cover afforded is warm and liked by the birds; and so for many reasons the fir has become a great favourite, notwithstanding that it is of very little value when finally cut down.

For fox preserving firs are hardly so suitable, because the needles, or small sharp leaves, quite destroy all under-growth—not only by the turpentine they contain, but by forming a thick mat, as it were, upon the earth. This mass of needles takes years, to all appearance, to decay, and no young green blade or shoot can get through it; besides which the fir-boughs above make a roof almost impenetrable to air and light, the chief necessities of a plant's existence. Foxes like a close warm under-growth, such as furze, sedges when the ground is dry, the under-

wood that springs up between the ash stoles. Although
constantly out of doors—if such a phrase be allowable—
foxes seem to dislike cold and draught, as do weasels and
all their kind, notably ferrets. But for pure game preserving, and for convenience of watching, the keeper thinks
the detached plantations of fir preferable. Doubtless he
is professionally right; and yet somehow a great wood
seems infinitely more English and appeals to the heart far
more powerfully, with its noble oaks and beeches and ash
trees, its bramble-thickets and brake, and endless beauties,
which a life of study will not exhaust.

But the semi-Bohemians detested by the keeper do
not prowl about the confines of a wood with artistic views;
their objects are extremely prosaic, and though not always
precisely injurious, yet they annoy him beyond endurance.
He is like a spider in the centre of a vast spreading web,
and the instant the most outlying threads—in this case
represented by fences—are broken he is all agitation till
he has expelled the intruder. Men and boys in the
winter come stealing into the wood where the blackthorn
thickets are for sloes, which are reputed to be improved
by the first frosts, and are used for making sloe gin, etc.
Those they gather they sell, of course; and although the
pursuit may be perfectly harmless in itself, how is the
keeper to be certain that, if opportunity offered, these
gentry would not pounce upon a rabbit or anything else?
Others come for the dead wood; and it does on the face

of it seem hard to deny an old woman who has worked
all her days in the field a bundle of fallen branches rotting
under the trees. The accumulations of such dead sticks
in some places are astonishing: the soil under the ashpoles must slowly rise from the mass of decaying wood
and ultimately become greatly enriched by this natural
manure.

When a hard clay soil is revealed by the operations
for draining a meadow, and the crust of black or reddish
mould on which the sweet green grass flourishes is seen to
be but spade-deep, the idea naturally occurs that that thin
crust must have been originated by some similar process
to what is now going on in the ash wood. Those six or
nine inches of mould perhaps represent several centuries of
forest. But if the keeper admits the old woman shivering
over her embers in the cottage to pick up these dead
boughs, how can he tell what further tricks others may
be up to? The privilege has often been offered and as
often abused, until at last it has been finally withdrawn—
not only because of the poaching carried on under the
cloak of picking up dead wood, but because the intruders
tore down fine living branches from the trees and spoiled
and disfigured them without mercy. Sometimes gentlemen go to the expense of having wood periodically
gathered and distributed among the poor, which is a con-
siderate system and worthy of imitation where possible.

Occasionally men come to search for walking-sticks,

for which there is now a regular trade. Just at present 'natural' sticks—that is, those cut from the stem with the bark on—are rather popular, both for walking and for umbrella handles, which causes this kind of search to be actively prosecuted. The best 'natural' sticks are those which when growing were themselves young trees, sprung up direct from seed or shoots—saplings, which are stronger and more pliant than those cut from a stole or pollard. To cut such a stick as this is equivalent to destroying a future tree, and of course a good deal of mischief may be easily done in a short time.

Another kind of ash stick which is in demand is one round which there runs a spiral groove. This spiral is caused by the bine of honeysuckle or woodbine, and in some cases by wild hops. These climbing plants grow in great profusion when they once get fixed in the soil, and twist their tendrils or 'leaders' round and round the tall, straight, young ash poles with so tight a grasp as to partly strangle the stick and form a deep screw-like groove in it. When well polished, or sometimes in its rough state, such a stick attract csustomers ; and so popular is this 'style' of thing that the spiral groove is frequently cut by the lathe in more expensive woods than ash. Wild hops are common in many places, and will almost destroy a hedge or a little copse by the power with which they twine their coils about stem and branch. Young oak saplings, in the same way, are frequently cut ; and the potential tree which might have grown large enough to form part of a ship's timbers is sold for a shilling.

Holly is another favourite wood for sticks, and fetches more money than oak or ash, on account of its ivory-like whiteness when peeled. To get a good stick with a knob to it frequently necessitates a considerable amount of cutting and chopping, and does far more damage than the loss of the stick itself represents. Neither blackthorn nor crabtree seem so popular as they once were for this purpose.

In the autumn scores of men, women, and children scour the hedges and woods for acorns, which bring a regular price per bushel or sack, affording a valuable food for pigs. Others seek elderberries to sell for making wine, and for a few weeks a trade is done in blackberries. Chair-menders and basket-makers frequent the shore of the little mere or lake looking for bulrushes or flags : the old rush-bottomed chairs are still to be found in country houses, and require mending ; and flag-baskets are much used.

Hazel-nuts and filberts perhaps cause more trouble than all the rest ; this fruit is now worth money, and in some counties the yield of nuts is looked forward to in the same way as any other crop—as in Kent, where cob-nuts are cultivated, and where the disorderly hop-pickers are great thieves. I have heard of owners of copses losing ten or fifteen pounds' worth of nuts by a single raid.

Here, in this wood, no attempt is made to obtain profit from the fruit, yet it gives rise to much trouble. The nut-stealers take no care in pulling down the boughs, but break them shamefully, destroying entire bushes ; and for this reason in many places, where nutting was once freely permitted, it is now rigidly repressed. Just before the nuts become ripe they are gathered by men employed on the place, and thrown down in sackfuls, making great heaps by the public footpaths—ocular evidence that it is useless to enter the wood a-nutting.

The keeper thinks that these trespassers grow more coarsely mischievous year by year. He can recollect when the wood in a measure was free and open, and, provided a man had not got a gun or was not suspected of poaching, he might roam pretty much at large ; while the resident labouring people went to and fro by the ·nearest short cut they could find. But whether the railways bring rude strangers with no respect for the local authorities, or whether ' tramps ' have become more numerous, it is certain that only by constant watchfulness can downright destruction be prevented. It is not only the game preserved within that closes these beautiful woodlands to the public, but the wanton damage to tree and shrub, the useless, objectless mischief so frequently practised. For instance, a column of smoke, curling like a huge snake round the limbs of a great tree and then floating away from the top-most branches, is a singular spectacle, so opposed to the ordinary current of ideas as to be certain of attracting the passer-by. It is the work, of course, of some mischievous lout who has set fire to the hollow interior of the tree.

Such a tree, as previously pointed out, is the favourite resort of bird and insect life. The heedless mischief of the bird-keeping boys, or the ploughlads rambling about on Sunday, destroys this Hôtel de Ville of the forest or hedgerow, the central house of assembly of the birds. To light a fire seems one of the special delights of these lads, and sometimes of men who should have learned better; and to light it in a hollow tree is the highest flight of genius. A few handfuls of withered grass and dead fern, half a dozen dry sticks, a lucifer-match, and the thing is done. The hollow within the tree is shaped like an inverted funnel, large at the bottom and decreasing upwards, where at the pointed roof one thin streak of daylight penetrates. This formation is admirably adapted to 'draw' a fire at the bottom, and so, once lit, it is not easily put out. The 'touchwood' smoulders and smokes immensely, and a great black column rises in the air. So it will go on smouldering and smoking for days till nothing but a charred stump be left. Now and then there is sufficient sap yet remaining in the bark and outer ring of wood to check the fire when it reaches it; and finally it dies out, being unable to burn the green casing of the trunk. Even then, so strong is the vital force, the oak may stand for years and put forth leaves on its branches

—leaves which, when dead, will linger, loth to fall, almost through the winter, rustling in the wind, till the buds of spring push them off.

Graver mischief is sometimes committed with the lucifer-match, and with more of the set purpose of destruction. In the vast expanse of furze outside the wood on the high ground the huntsmen are almost certain of a find, and, if they can get between the fox and the wood, of a rattling burst along the edge of the downs ; no wonder, therefore, that both they and the keeper set store by this breadth of 'bush.' To this great covert more than once

AN ENGLISH PRAIRIE-FIRE.

some skulking scoundrel has set fire, taking good care to strike his match well to windward, so that the flames might drive across the whole, and to choose a wind which would also endanger the wood. Now nothing flares up with

such a sudden fierceness as furze, and there is no possi-
bility of stopping it.   With a loud crackling, and swaying
of pointed tongues of flame visible miles away even at
noontide, and a cloud of smoke, the rift rolls on, licking
up grass and fern and heath ; and its hot breath goes
before it, and the blast rises behind it.   As on the beach
the wave seems to break at the foot, and then in an instant
the surf runs away along the sand, so from its first start
the flame widens out right and left with a greedy eagerness,
and what five minutes ago was but a rolling bonfire is now
a wall of fire a quarter of a mile broad, and swelling as it
goes.

Then happens on a lesser scale exactly the same thing
that travellers describe of the burning prairies of the Far
West—a stampede of the thousands of living creatures,
bird and beast : rabbits, hares, foxes, weasels, stoats,
badgers, wild cats, all rushing in a maddened frenzy of
fear they know not whither.   Often, with a strange
reversal of instinct, so to say, they will crowd together
right in the way of the flames, huddling in hundreds where
the fire must pass, and no effort of voice or presence of
man will drive them away.   The hissing, crackling fire
sweeps over, and in an instant all have perished.   No
more miserable spectacle can be witnessed than the terror
of these wretched creatures.   Birds seem to fly into the
smoke and are suffocated—they fall and are burned.
Hares, utterly beside themselves, will rush almost into the

arms of the crowd that assembles, and, of course, picks up what it can seize.   The flames blacken and scorch the firs and trees on the edge of the wood, and the marks of their passage are not obliterated for years.

Apart from the torture of animals, the damage to sport—both hunting and shooting—is immense, and takes long to remedy ; for although furze and fern soon shoot again, yet animal life is not so quickly repaired.   Sometimes a few sheep wandering from the downs are roasted alive in this manner ; and one or more dogs from the crowd watching are sure to run into the flames, which seem to exercise a fascination over some canine minds. The keeper's wrath bubbles up years afterwards as he recalls the scene, and it would not be well for the incendiary if he fell into his hands.   But the mischief can be so easily done that it is rarely these rascals are captured.

CHAPTER VII.

Professional Poachers.—The art of Wiring Game.

THERE are three kinds of poachers, the local men, the raiders coming in gangs from a distance, and the 'mouchers'—fellows who do not make precisely a profession of it, but who occasionally loiter along the roads and hedges picking up whatever they can lay hands on. Philologists may trace a resemblance between the present provincial word 'mouching' and Shakspeare's 'mitcher,' who ate blackberries. Of the three probably the largest amount of business is done by the local men, on the principle that the sitting gamester sweeps the board. They therefore deserve first consideration.

It is a popular belief that the village poacher is an idle, hang-dog ne'er-do-well, with a spice of sneaking romance in his disposition—the Bohemian of the hamlet, whose grain of genius has sprouted under difficulties, and produced weeds instead of wheat. This is a complete fallacy, in our day at least. Poaching is no longer an amusement, a thing to be indulged in because

> It's my delight of a shiny night
> In the season of the year ;

but a hard, prosaic business, a matter of £ s. d., requiring a long-headed, shrewd fellow, with a power of silence, capable of a delicacy of touch which almost raises poaching into a fine art. The real man is often a sober and to all appearance industrious individual, working steadily during the day at some handicraft in the village, as blacksmithing, hedge-carpentering—i.e. making posts and rails, etc.—cobbling, tinkering, or perhaps in the mill; a somewhat reserved, solitary workman of superior intelligence and frequently advanced views as to the 'rights of labour.' He has no appetite for thrilling adventure; his idea is simply money, and he looks upon his night-work precisely as he does upon his day-labour.

His great object is to avoid suspicion, knowing that success will be proportionate to his skill in cloaking his operations; for in a small community, when a man is 'suspect,' it is comparatively easy to watch him, and a poacher knows that if he is watched he must sooner or later be caught. Secrecy is not so very difficult; for it is only with certain classes that he need practise concealment: his own class will hold their peace. If a man is seen at his work in the day, if he is moderate in his public-house attendance, shows himself at church, and makes friends with the resident policemen (not as a confederate, but to know his beat and movements), he may go on for years without detection.

Perhaps the most promising position for a man who

L

makes a science of it is a village at the edge of a range of
downs, generally fringed with large woods on the lower
slopes.   He has then ground to work alternately, according to the character of the weather and the changes of the
moon.   If the weather be wet, windy, or dark from the
absence of the moon, then the wide open hills are safe ;
while, on the other hand, the woods are practically in-
accessible, for a man must have the eyes of a cat to see to
do his work in the impenetrable blackness of the plantations.   So that upon a bright night the judicious poacher
prefers the woods, because he can see his way, and avoids
the hills, because, having no fences to speak of, a watcher
may detect him a mile off.

Meadows with double mounds and thick hedges may
be worked almost at any time, as one side of the hedge is
sure to cast a shadow, and instant cover is afforded by the
bushes and ditches.   Such meadows are the happy hunting-
grounds of the poacher for that reason, especially if not
far distant from woods, and consequently overrun with
rabbits.   For, since the price of rabbits has risen so high,
they are very profitable as game, considering that a dozen
or two may be captured without noise and without having
to traverse much space—perhaps in a single hedge.

The weather most unsuitable is that kind of frost which
comes on in the early morning, and is accompanied with
some rime on the grass—a duck's frost, just sufficient to
check fox-hunting.   Every footstep on grass in this con-

dition when the sun comes out burns up as black as if the
sole of the boot were of red-hot iron, and the poacher
leaves an indelible trail behind him.   But as three duck's
frosts usually bring rain, a little patience is alone necessary.
A real, downright six weeks' frost is, on the contrary, very
useful—game lie close.   But a deep snow is not welcome ;
for, although many starved animals may be picked up, yet
it quite suspends the operations of the regular hand : he
can neither use wire, net, nor ferret.

Windy nights are disliked, particularly by rabbit-
catchers, who have to depend a great deal upon their sense
of hearing to know when a rabbit is moving in the 'buries,'
and where he is likely to ' bolt,' so as to lay hands on him
the instant he is in the net.   But with the 'oak's mysterious
roar ' overhead, the snapping of dead branches, and the
moan of the gale as it rushes through the hawthorn, it is
difficult to distinguish the low, peculiar thumping sound of
a rabbit in his catacomb.   The rabbit is not easily dis-
lodged in rain ; for this animal avoids getting wet as much
as possible : he ' bolts ' best when it is dry and still.

A judicious man rarely uses a gun, for the reason that
noise is inconvenient, and a gun is an awkward tool to
carry concealed about the person even when taken to
pieces.   There is a certain prejudice in rural places against
a labouring man possessing a gun ; it is sure to draw
suspicion upon him.   A professional poacher is pre-
eminently a trapper, relying chiefly upon the dexterous

employment of the snare.   If he does shoot, by preference
he chooses a misty day, knowing that the sound of the
report travels scarcely half the usual distance through fog;
and he beats the meadows rather than the preserves, where
the discharge would instantly attract attention, while in the
meadows or ploughed fields it may pass unnoticed as fired
by a farmer with leave to kill rabbits.

PARTRIDGES AT EVENING.

When the acorns are ripe and the pheasants wander
great distances from the plantations along the hedgerows is

his best time for shooting; no keepers at that period can protect them. He also observes where the partridges which roost on the ground assemble nightly as it grows dark, easily ascertaining the spot by their repeated calls to each other, and sometimes knocks over three or four at a shot.

Occasionally, also, early in the season, before the legitimate sportsman perhaps has stepped into the stubble, and while the coveys are large, he sees a good chance, and with two or even three ounces of shot makes havoc among them. He invariably fires at his game sitting, first, because he cannot lose an opportunity, and, next, because he can kill several at once. He creeps up behind a hedge, much as the sportsman in Rubens' picture in the National Gallery is represented, stooping to get a view, himself unseen, at the brown birds on the ground. With the antique firelock such a practice was necessary; but nothing in our day so stamps a man a poacher as this total denial of 'law' to the game.

When the pheasant is shot his next difficulty is with the feathers. The fluffy, downy under-feathers fly in all directions, scattering over the grass, and if left behind would tell an unmistakable tale. They must therefore be collected as far as possible, and hidden in the ditch. The best pockets for carrying game are those made in the tails of the coat, underneath: many poachers' coats are one vast pocket behind the lining.

When there is special danger of being personally over-

hauled and searched, or when the 'bag' is large, the game
is frequently hidden in a rabbit-hole, taking care to fence
the hole some distance inside with a stout stick across it,
the object of which is that if the keeper or a sportsman

RUBENS' SPORTSMAN.

should pass that way his dogs, scenting the game, will
endeavour to scratch out the earth and get in after it.
This the cross stick will prevent, and the keeper will probably thrash his dog for refusing to obey when called off.

A great deal of poaching used to be accomplished by

nets, into which both partridges and pheasants were driven. If skilfully alarmed—that is, not too much hurried—these birds will run a long way before rising, and, if their tracks are known, may be netted in considerable numbers. But of recent years, since pheasants especially have become so costly a luxury to keep, the preserves and roosting-places have been more effectually watched, and this plan has become more difficult to put in practice. In fact, the local man thinks twice before he puts his foot inside a preserve, and, if possible, prefers to pick up outside. If a preserve is broken into the birds are at once missed, and there is a hue and cry; but the loss of outsiders is not immediately noticed.

The wire is, perhaps, the regular poacher's best implement, and ground game his most profitable source of income. Hares exist in numbers upon the downs, especially near the localities where the great coursing meetings are held, where a dozen may be kicked out of the grass in five minutes. In these districts of course the downs are watched; but hares cannot be kept within bounds, and wander miles and miles at night, limping daintily with their odd gait (when undisturbed) along the lanes leading into the ploughed fields on the lower slopes and plains. The hills—wide and almost pathless, and practically destitute of fences—where the foot leaves no trail on the short grass and elastic turf, are peculiarly favourable to illicit sport.

Though apparently roaming aimlessly, hares have their
regular highways or 'runs ;' and it is the poacher's business
to discover which of these narrow paths are most beaten by
continuous use.   He then sets his wire, as early in the
evening as compatible with safety to himself, for hares are
abroad with the twilight.

Long practice and delicate skill are essential to successful snaring.   First, the loop itself into which the hare is to
run his head must be of the exact size.   If it be too small
he will simply thrust it aside ; if too large his body will
slip through, and his hind leg will be captured : being
crooked, it draws the noose probably.   Then if caught by
the hind leg, the wretched creature, mad with terror, will
shriek his loudest ; and a hare shrieks precisely like a
human being in distress.   The sound, well understood by
the watchers, will at once reveal what is going forward.
But there may be no watchers about ; and in that case
the miserable animal will tug and tug during the night
till the wire completely bares the lower bone of the leg,
and in the morning, should any one pass, his leaps and
bounds and rolls will of course be seen.   Sometimes he
twists the wire till it snaps, and so escapes—but probably
to die a lingering death, since the copper or brass is pretty
sure to mortify the flesh.   No greater cruelty can be imagined.   The poacher, however, is very anxious to avoid
it, as it may lead to detection ; and if his wire is properly
set the animal simply hangs himself, brought up with a

sudden jerk which kills him in two seconds, and with less pain than is caused by the sting of the sportsman's cartridge.

SETTING A HARE SNARE.

Experience is required to set the loop at the right height above the ground. It is measured by placing the

clenched fist on the earth, and then putting the extended
thumb of the other open hand upon it, stretching it out as
in the action of spanning, when the tip of the little finger
gives the right height for the lower bend of the loop—
that is, as a rule ; but clever poachers vary it slightly to
suit the conformation of the ground.   A hare carries his
head much higher than might be thought ; and he is very
strong, so that the plug which holds the wire must be
driven in firmly to withstand his first convulsive struggle.
The small upright stick whose cleft suspends the wire
across the 'run' must not be put too near the hare's path,
or he will see it, and it must be tolerably stiff, or his head
will push the wire aside.   Just behind a 'tussocky' bunch
of grass is a favourite spot to set a noose ; the grass partially conceals it.

The poacher revisits his snares very early in the
morning, and if he is judicious, invariably pulls them up,
whether successful or not, because they may be seen in
the day.   Half the men who are fined by the magistrates
have been caught by keepers who, having observed wires,
let them remain, but keep a watch and take the offenders
red-handed.   The professional poacher never leaves his
wires set up all day, unless a sudden change of weather
and the duck's frost previously mentioned prevent him
from approaching them, and then he abandons those
particular snares for ever.   For this reason he does not
set up more than he can easily manage.   If he gets three

hares a night (wholesale price 2s. 6d. each) he is well repaid.   Rabbits are also wired in great numbers.   The loop is a trifle smaller, and should be just a span from the ground.

But the ferret is the poacher's chief assistant in rabbiting : it takes two men, one on each side of the 'bury,' and a ferret which will not 'lie in '—*i.e.* stay in the hole and feast till overcome with sleep.   Ferrets differ remarkably in disposition, and the poacher chooses his with care ; otherwise, if the ferret will not come out, the keepers are certain to find him the next day hunting on his own account.   Part of the secret is to feed him properly, so that he may have sufficient appetite to hunt well and yet be quickly satisfied with a taste of blood.   Skill is essential in setting up the nets at the mouth of the holes ; but beyond the mere knack, easily acquired, there is little to learn in ferreting.

The greatest difficulty with any kind of game is to get home unobserved with the bag.   Keepers are quite aware of this ; and in the case of large estates, leaving one or two assistants near the preserves, they patrol the byways and footpaths, while the police watch the crossroads and lanes which lead to the villages.   If a man comes along at an exceptionally early hour with coat pockets violently bulging, there is a *prima facie* case for searching him.   One advantage of wiring or netting over the gun is here very noticeable : anything shot bleeds and stains the pocket—a suspicious sign even when empty; strangulation leaves no traces.

Without a knowledge of the policeman's beat and the keeper's post the poacher can do nothing on a large scale. He has, however, no great trouble in ascertaining these things ; the labourers who do not themselves poach sympathise warmly and whisper information. There is reason to think that men sometimes get drunk, or sufficiently so to simulate intoxication very successfully, with the express purpose of being out all night with a good excuse, and so discovering the policeman's ambuscade. Finding a man, whom he knows to be usually sober, overtaken with drink in a lonely road, where he injures none but himself, the policeman goodnaturedly leads him home with a caution only.

The receivers of game are many and various. The low beer-shop keepers are known to purchase large quantities. Sometimes a local pork-butcher in a small way buys and transmits it, having facilities for sending hampers, etc., unsuspected. Sometimes the carriers are the channel of communication ; and there is no doubt the lower class of game dealers in the provincial towns get a good deal in this way. The London dealer, who receives large consignments at once, has of course no means of distinguishing poached from other game. The men who purchase the rabbits ferreted by the keepers during the winter in the woods and preserves, and who often buy

£100 worth or more in the season, have peculiar opportunities for conveying poached animals, carefully stowed
for them in a ditch on their route. This fact having crept
out has induced gentlemen to remove these rabbit contracts from local men, and to prefer purchasers from a
distance, who must take some time to get acquainted with
the district poachers.

The raiders, who come in gangs armed with guns and
shoot in the preserves, are usually the scum of manufacturing towns, led or guided by a man expelled through
his own bad conduct from the village, and who has a
knowledge of the ground. These gangs display no skill ;
relying on their numbers, arms, and known desperation of
character to protect them from arrest, as it does in nine
cases out of ten. Keepers and policemen cannot be expected to face such brutes as these fellows ; they do sometimes, however, and get shattered with shot.

The 'mouchers' sneak about the hedgerows on Sun-
days with lurcher dogs, and snap up a rabbit or a hare ;
they do not do much damage except near great towns,
where they are very numerous. Shepherds, also, occasionally mouch—their dogs being sometimes very expert ;
and ploughmen set wires in the gateways or gaps where
they have noticed the track of a hare, but it is only for
their own eating, and is not of much consequence in comparison with the work of the real local professional.
These regular hands form a class which are probably more

numerous now than ever ; the reasons are—first, the high
value of game and the immense demand for it since
poultry has become so dear, and, secondly, the ease of
transmission now that railways spread into the most outlying districts and carry baskets or parcels swiftly out of
reach.   Poaching, in fact, well followed is a lucrative
business.

Some occasional poaching is done with no aid but the
hand, especially in severe weather, which makes all wild
animals 'dummel,' in provincial phrase—*i.e.* stupid, slow
to move.   Even the hare is sometimes caught by hand as
he crouches in his form.   It requires a practised eye, that
knows precisely where to look among the grass, to detect
him hidden in the bunch under the dead, dry bennets.
An inexperienced person chancing to see a hare sitting
like this would naturally stop short in walking to get a
better view ; whereupon the animal, feeling that he was
observed, would instantly make a rush.   You must
persuade the hare that he is unseen ; and so long as he
notices no start or sign of recognition—his eye is on you
from first entering the field—he will remain still, believing
that you will pass.

The poacher, having marked his game, looks steadily
in front of him, never turning his head, but insensibly
changes his course and quietly approaches sidelong.   Then,
in the moment of passing, he falls quick as lightning on
his knee, and seizes the hare just behind the poll.   It is

POACHING IN THE WINTER.

the only place where the sudden grasp would hold him in
his convulsive terror—he is surprisingly powerful—and
almost ere he can shriek (as he will do) the left hand has
tightened round the hind legs. Stretching him to his full
length across the knee, the right thumb, with a peculiar
twist, dislocates his neck, and he is dead in an instant.
There is something of the hangman's knack in this, which
is the invariable way of killing rabbits when ferreted or
caught alive ; and yet it is the most merciful, for death

is instantaneous.   It is very easy to sprain the thumb
while learning the trick.

A poacher will sometimes place his hat gently on
the ground, when first catching sight of a sitting hare, and
then stealthily approach on the opposite side.   The hare
watches the hat, while the real enemy comes up unawares,
or, if both are seen, he is in doubt which way to dash.
On a dull, cold day hares will sit till the sportsman's dogs
are nearly on them, almost till he has to kick them out.
At other times in the same locality they are, on the contrary, too wild.   Occasionally a labourer, perhaps a 'fogger,'
crossing the meadows with slow steps, finds a rabbit sitting
in like manner among the grass or in a dry furrow.
Instantly he throws himself all a-sprawl upon the ground,
with the hope of pinning the animal to the earth.   The
manœuvre, however, frequently fails, and the rabbit slips
away out of his very hands.

The poacher is never at rest ; there is no season when
his marauding expeditions cease for awhile : he acknow-
ledges no 'close time' whatever.   Almost every month
has its appropriate game for him, and he can always turn
his hand to something.   In the very heat of the summer
there are the young rabbits, for which there is always a
sale in the towns, and the leverets, which are easily picked
up by a lurcher dog.

I have known a couple of men take a pony and trap
for this special purpose, and make a pleasant excursion

over hill and dale, through the deep country lanes, and
across the open down land, carrying with them two or
three such dogs to let loose as opportunity offers.   Their
appearance as they rattle along is certainly not prepossess-
ing ; the expression of their canine friends trotting under
the trap, or peering over the side, stamps them at the first
glance as ' snappers up of unconsidered trifles ; ' but you
cannot arrest these gentlemen peacefully driving on the
' king's highway ' simply because they have an ugly look
about them.   From the trap they get a better view than
on foot ; standing up they can see over a moderately high
hedge, and they can beat a rapid retreat if necessary, with
the aid of a wiry pony.   Passing by some meadows, they
note a goodly number of rabbits feeding in the short
aftermath.   They draw up by a gateway, and one of them
dismounts.   With the dogs he creeps along behind the
hedge (the object being to get between the bunnies and
their holes), and presently sends the dogs on their mission.
The lurchers are tolerably sure of catching a couple—
young rabbits are neither so swift nor so quick at doubling
as the older ones.   Before the farmer and his men, who
are carting the summer-ricks in an adjacent field, can quite
comprehend what the unusual stir is about yonder, the
poachers are off, jogging comfortably along, with their game
hidden under an old sack or some straw.

Their next essay is among the ploughed fields, where
the corn is ripening and as yet no reapers are at work, so

M

that the coast is almost clear. Here they pick up a leveret, and perhaps the dogs chop a weakly young partridge, unable to fly well, in the hedge. The keeper has just strolled through the copses bordering on the road and has left them, as he thinks, safe. They watch his figure slowly disappearing in the distance from a bend of the lane, and then send the dogs among the underwood. In the winter men will carry ferrets with them in a trap like this.

The desperate gangs who occasionally sweep the preserves, defying the keepers in their strength of numbers and prestige of violence, sometimes bring with them a horse and cart, not so much for speed of escape as to transport a heavy bag of game. Such a vehicle, driven by one man, will, moreover, often excite no suspicion, though it may be filled with pheasants under sacks and hay. A good deal of what may be called casual poaching is also done on wheels.

Some of the landlords of the low beer-houses in the country often combine with the liquor trade the business of dealing in pigs, calves, potatoes, etc., and keep a light cart, or similar conveyance. Now, if any one will notice the more disreputable of these beer-houses, they will observe that there are generally a lot of unkempt, rough-looking dogs about them. These, of course, follow their master when he goes on his short journeys from place to place ; and they are quite capable of mischief. Such men may not make a business of poaching, yet if in passing a preserve the dogs stray and bring back something eatable, why, it is very easy to stow it under the seat with the potatoes.   Sometimes a man is bold enough to carry a gun in this way—to jump out when he sees a chance and have a shot, and back and off before any one knows exactly what is going on.

Somehow there always seems to be a market for game out of season : it is 'passed' somewhere, just as thieves pass stolen jewellery.   So also fish, even when manifestly unfit for table, in the midst of spawning time, commands a ready sale if overlooked by the authorities.   It is curious that people can be found to purchase fish in such a condition ; but it is certain that they do.   In the spring, when one would think bird and beast might be permitted a breathing space, the poacher is as busy as ever after eggs.   Pheasant and partridge eggs are largely bought and sold in the most nefarious manner.   It is suspected that some of the less respectable breeders who rear game birds like poultry for sale, are not too particular of whom they purchase eggs ; and, as we have before observed, certain keepers are to blame in this matter also.

Plovers' eggs, again, are an article of commerce in the spring ; they are protected now by law, but it is to be feared that the enactment is to a great extent a dead letter.   The eggs of the peewit, or lapwing, as the bird is variously called, are sought for with great perseverance, and accounted delicacies.   These birds frequent commons

where the grass is very rough, and interspersed with
bunches of rushes, marshy places, and meadows liable to
be flooded in the winter. The nest on the ground is often
made in the depression left by a horse's hoof in the soft

PLOVER'S NEST.

earth—any slight hole, in fact ; and it is so concealed, or
rather differs so little from the appearance of the general
sward around, as to be easily passed unnoticed. You
may actually step on it, and so smash the eggs, before you
see it.

Aware that the most careful observation may fail to

find what he wants, the egg-stealer adopts a simple but effective plan by which he ensures against omitting to examine a single foot of the field. Drop a pocket-knife or some such object in the midst of a great meadow, and you will find the utmost difficulty in discovering it again, when the grass is growing tall as in spring. You may think that you have traversed every inch, yet it is certain that you have not; the inequalities of the ground insensibly divert your footsteps, and it is very difficult to keep a straight line. What is required is something to fix the eye—what a sailor would call a 'bearing.' This the egg-stealer finds in a walking-stick. He thrusts the point into the earth, and then slowly walks round and round it, enlarging the circle every time, and thus sweeps every inch of the surface with his eye. When he has got so far from the stick as to feel that his steps are becoming uncertain he removes it, and begins again in another spot. A person not aware of this simple trick will search a field till weary and declare there is nothing to be found; another, who knows the dodge, will go out and return in an hour with a pocketful of eggs.

On those clear, bright winter nights when the full moon is almost at the zenith, and the 'definition' of tree and bough in the flood of light seems to equal if not to exceed that of the noonday, some poaching used to be accomplished with the aid of a horsehair noose on the end of a long slender wand. There are still some districts in

the country more or less covered with forest, and which
on account of ancient rights cannot be enclosed.   Here
the art of noosing lingers ; the loop being insidiously
slipped over the bird's head while at roost.   By constant
practice a wonderful dexterity may be acquired in this
trick ; men will snare almost any bird in broad daylight.
With many birds a favourite place for a nest is in a
hollow tree, access being had by a decayed knothole, and
they are sometimes noosed as they emerge.   A thin
flexible copper wire is said to be substituted for large
game.   This method of capture peculiarly suits the views
of the ornithologist, with whom it is an object to avoid
the spoiling of feathers by shot.

Every now and then a bird-catcher comes along
decoying the finches from the hedges, for sale as cage-
birds in London.   Some of these men, without any
mechanical assistance, can imitate the 'call' note of the
bird they desire to capture so as to deceive the most
practised ear.   These fellows are a great nuisance, and
will completely sweep a lane of all the birds whose song
makes them valuable.   In this way some localities have
been quite cleared of goldfinches, which used to be common.   The keepers, of course, will not permit them on
private property ; but in all rural districts there are wide
waste spaces—as where two or more roads meet—broad
bands of green sward running beside the highway, and
the remnants of what in former days were commons ; and

here the bird-catcher plies his trade. It so happens that these very waste places are often the most favourite resorts of goldfinches, for instance, who are particularly fond of thistledown, and thistles naturally chiefly flourish on uncultivated land. These men and the general class of loafers have a wholesome dread of gamekeepers, who look on them with extreme suspicion.

The farmers and rural community at large hardly give the gamekeeper his due as a protection against thieves and mischievous rascals. The knowledge that he may at any time come round the corner, even in the middle of the night, has a decidedly salutary effect upon the minds of those who are prowling about. Intoxicated louts think it fine fun to unhinge gates, and let cattle and horses stray abroad, to tear down rails, and especially to push the coping stones off the parapets of the bridges which span small streams. They consider it clever to heave these over with a splash into the water, or to throw down half a dozen yards of 'dry wall.' In many places fields are commonly enclosed by the roadside with such walls, which are built of a flat stone dug just beneath the surface, and used without mortar. There are men who make a business of building these walls; it requires some skill and patience to select the stones and fit them properly. They serve the purpose very well, but the worst is that if once started the process of destruction is easy and quick. Much more serious offences than these are sometimes committed, as cutting horses with knives, and other mutilations. The fact that the fields are regularly perambulated by keepers and their assistants night and day cannot but act as a check upon acts of this kind.

CHAPTER VIII.

The Field Detective—Fish Poaching.

THE footpaths through the plantations and across the fields have no milestones by which the pedestrian can calculate the distance traversed ; nor is the time occupied a safe criterion, because of the varying nature of the soil —now firm and now slippery—so that the pace is not regular. But these crooked paths—no footpath is ever straight—really represent a much greater distance than would be supposed if the space from point to point were measured on a map. So that the keeper as he goes his rounds, though he does not rival the professional walker, in the course of a year covers some thousands of miles. He rarely does less than ten, and probably often twelve miles a day, visiting certain points twice — *i.e.* in the morning and evening—and often in addition, if he has any suspicions, making *détours*. It is easy to walk a mile in a single field of no great dimensions when it is necessary to go up and down each side of four long hedgerows, and backwards and forwards, following the course of the furrows.

The keeper's eye is ever on the alert for the poacher's

wires ; and where the grass is tall to discover these is often a tedious task, since he may go within a few yards and yet pass them. The ditches and the great bramble-bushes are carefully scanned, because in these the poacher often conceals his gun, nets, or game, even when not under immediate apprehension of capture. The reason is that his cottage may perhaps be suddenly searched : if not by authority, the policeman on some pretext or other may unexpectedly lift the latch or peer into the outhouses, and feathers and fur are apt to betray their presence in the most unexpected manner. One single feather, one single fluffy little piece of fur overlooked, is enough to ruin him, for these are things of which it is impossible to give an acceptable explanation.

In dry weather the poacher often hides his implements; especially is this the case after a more than usually venture-some foray, when he knows that his house is tolerably certain to be overhauled and all his motions watched. A hollow tree is a common resource—the pollard willow generally becomes hollow in its old age—and with a piece of the decaying ' touchwood ' or a strip of dead bark his tools are ingeniously covered up. Under the eaves of sheds and outhouses the sparrows make holes by pulling out the thatch, and roost in these sheltered places in severe weather, warmly protected from the frost ; other small birds, as wrens and tomtits, do the same ; and the poacher avails himself of these holes to hide his wires.

A gun has been found before now concealed in a heap of manure, such as are frequently seen in the corners of the fields.   These heaps sometimes remain for a year or more in order that the materials may become thoroughly decomposed, and the surface is quickly covered with a rank growth of weeds.   The poacher, choosing the side close to the hedge, where no one would be likely to go, excavated a place beneath these weeds, partly filled it in with dry straw, and laid his gun on this.   A rough board placed over it shielded it from damp ; and the aperture was closed with ' bull-polls '—that is, the rough grass of the furrows chopped up (not unlike the gardener's ' turves ') —and thrown on the manure-heap to decay.   If the keeper detects anything of this kind he allows it to stay undisturbed, but sets a watch, and so surprises the owner of the treasure.

The keeper is particularly careful to observe the motions of the labourers engaged in the fields ; especially at luncheon-time, when men with a hunch of bread and a slice of bacon—kept on the bread with a small thumb-piece of crust, and carved with a pocket knife—are apt to ramble round the hedges, of course with the most innocent of motives, admiring the beauties of nature. Slowly wandering like this, they cast a sidelong glance at their wire, set up in a ' drock '—*i.e.* a bridge over a ditch formed of a broad flat stone—which chances to be a favourite highway of the rabbits.   Nowadays, in this age

of draining, short barrel drains of brick or large glazed
pipes are often let through thick banks ; these are dry for
weeks together, and hares slip through them.   A wire or
trap set here is quite out of ordinary observation ; and
the keeper, who knows that he cannot examine every inch
of ground, simplifies the process by quietly noting the
movements of the men.   As he passes and repasses a
field where they are at work day after day, and understands
agricultural labour, he is aware that they have no necessity
to visit hedgerows and mounds a hundred yards distant,
and should he see anything of that kind the circle of his
suspicions gradually narrows till he hits the exact spot
and person.

The gateways and gaps receive careful attention—
unusual footmarks in the mud are looked for.   Sometimes
he detects a trace of fur or feathers, or a bloodstain on
the spars or rails, where a load of rabbits or game has
been hung for a few minutes while the bearer rested.   The
rabbit-holes in the banks are noted : this becomes so much
a matter of habit as to be done almost unconsciously and
without effort as he walks ; and anything unusual—as the
sand much disturbed, the imprint of a boot, the bushes
broken or cut away for convenience of setting a net—is seen
in an instant.   If there be any high ground—woods are often
on a slope—the keeper has here a post whence to obtain
a comprehensive survey, and he makes frequent use of this
natural observatory, concealing himself behind a tree trunk.

A RABBIT-HOLE NETTED.

The lanes and roads and public footpaths that cross
the estate near the preserves are a constant source of un-
easiness. Many fields are traversed by a perfect network
of footpaths, half of which are of very little use, but cannot be closed. Nothing causes so much ill-will in rural
districts as the attempt to divert or shut up a track like
this. Cottagers are most tenacious of these 'rights,' and
will rarely exchange them for any advantage. 'There
always wur a path athwert thuck mead in the ould volk's
time' is their one reply endlessly reiterated; and the
owner of the property, rather than make himself unpopular,
desists from persuasion. The danger to game from these

paths arises from the impossibility of stopping a suspicious character at once.   If he breaks through a hedge it is different ; but the law is justly jealous of the subject's liberty on a public footpath, and you cannot turn him back.

Neither is it of any use to search a man whose tools, to a moral certainty, are concealed in some hedge.   With his hands in his pockets, and a short pipe in his mouth, he can saunter along the side of a preserve if only a path, as is often the case, follows the edge ; and by-and-by it grows dusk, and the keeper or keepers cannot be every-where at once.   There is nothing to prevent such fellows as these from sneaking over an estate with a lurcher dog at their heels—a kind of dog gifted with great sagacity, nearly as swift as a greyhound, and much better adapted for picking up the game when overtaken, which is the greyhound's difficulty.   They can be taught to obey the faintest sign or sound from their owners.   If the latter imagine watchers to be about, the lurcher slinks along close behind, keeping strictly to the path.   Presently, if the poacher but lifts his finger, away dashes the dog, and will miss nothing he comes across.   The lurcher has always borne an evil repute, which of course is the due not of the dog but of his master.   If a man had to get his living by the chase in Red Indian fashion, probably this would be the best breed for his purpose.   Many shepherds' dogs now have a cross of the lurcher in their strain, and are good at poaching.

A WICKED LURCHER AND HIS MASTER.

Sunday is the gamekeeper's worst day; the idle, rough characters from the adjacent town pour out into the country, and necessitate extra watchfulness. On Sunday, the keeper, out of respect to the day, does not indeed carry his gun, but he works yet harder than on week-days. While the chimes are ringing to church he is on foot by the edge of the preserves. He has to maintain a sort of surveillance over the beer-houses in the village, which is done with the aid of the district policeman, for they are not only the places where much of the game is sold, but the rendezvous of those who are planning a raid.

If the policeman notices an unusual stir, or the arrival of strange men without any apparent business, he acquaints the keeper, who then takes care to double his sentinels, and personally visit them during the night. This night-work is very trying after his long walks by day. A great object is to be about early in the morning—just before the dawn ; that is the time when the poachers return to examine their wires. By day he often varies his rounds so as to appear upon the scene when least expected ; and has regular trysting places, where his assistants meet him with their reports.

The gipsies, who travel the road in caravans, give him endless trouble ; they are adepts at poaching, and each van is usually accompanied by a couple of dogs. The movements of these people are so irregular that it is impossible to be always ready for them. They are suspected of being recipients of poached game, purchasing it from the local professionals. Under pretence of cutting skewer-wood, often called dogwood, which they split and sharpen for the butchers, they wander across the open downs where it grows, camping in wild, unfrequented places, and finding plenty of opportunities for poaching. Down land is most difficult to watch.

Then the men who come out from the towns, ostensibly

to gather primroses in the early spring, or ferns, which they hawk from door to door ; and the watercress men, who are about the meadows and brooks twice a year, in spring and autumn, require constant supervision. An innocent-looking basket or small sack-bag of mushrooms has before now, when turned upside down, been discovered to contain a couple of rabbits or a fine young leveret. This detective work is, in fact, never finished. There is no end to the tricks and subterfuges practised, and with all his experience and care the gamekeeper is frequently outwitted.

The relations between the agricultural labourers and the keeper are not of the most cordial character ; in fact, there is a ceaseless distrust upon the one hand and incessant attempts at over-reaching upon the other. The ploughmen, the carters, shepherds, and foggers, have so many opportunities as they go about the fields, and they never miss the chance of a good dinner or half-a-crown when presented to them. Higher wages have not in the slightest degree diminished poaching, regular or occasional ; on the contrary, from whatever cause, there is good reason to believe it on the increase. If a labourer crossing a field sees a hare or rabbit crouching in his form, what is to prevent him from thrusting his prong like a spear suddenly through the animal and pinning him to the turf ? There are plenty of ways of hiding dead game, under straw or hay, in the thick beds of nettles which usually spring up outside or at the back of a cowshed.

N

Why does the keeper take such a benevolent interest
in the progress of spade-husbandry, as exemplified in
allotment gardens near the village, which allotments are

THE GENTLEMAN IN VELVETEEN.

generally in a field set apart by the principal landowner
for the purpose?   In person or by proxy the keeper is very
frequently seen looking over the close-cropped hedge which

surrounds the spot, and now and then he takes a walk up and down the narrow paths between the plots. His dog sniffs about among the heaps of rubbish or under the potato-vines. The men at work are remarkably civil and courteous to the gentleman in the velveteen jacket, who on his side, is equally chatty with them ; but both in their hearts know very well the why and wherefore of this interest in agriculture. Almost all kinds of game are attracted by gardens, presupposing, of course, that they are situated at a distance from houses, as these allotments are. There is a supply of fresh, succulent food of various kinds : often too, after a large plot has been worked for garden produce, the tenant will sow it for barley or beans or oats, on the principle of rotation ; and these small areas of grain have a singular fascination for pheasants, and hares linger in them.

Rabbits, if undisturbed, are particularly fond of garden vegetables. In spring and early summer they will make those short holes in which they bring forth their young under the potato-vines, finding the soil easy to work, dry, and the spot sheltered by the thick green stems and leaves. Both rabbits and hares do considerable damage if they are permitted the run of the place unchecked. The tenants of the allotments, however, instead of driving them off, are anxious that they should come sniffing and limping over the plots in the gloaming, and are strongly suspected of allowing crops specially pleasing to game to remain in the

ground till the very latest period in order that they may
snare it.

Much kindly talk has been uttered over allotments,
and undoubtedly they are a great encouragement to the
labourer ; yet even this advantage is commonly abused.
The tenants have no ground of complaint as to damage to
their crops, because the keeper, at a word from them, would
lose not a moment of time in killing or driving away the
intruder ; and as an acknowledgment of honesty and in
reparation of the mischief, if any, a couple of rabbits
would be presented to the man who carried the complaint.
But the labourer, if he spies the tracks of a hare running
into his plot of corn, or suspects that a pheasant is hiding
there, carefully keeps that knowledge to himself.   He
knows that a pheasant, if you can get close enough to it
before it rises, is a clumsy bird, and large enough to offer
a fair mark, and may be brought down with a stout stick
dexterously thrown.   As very probably the pheasant is
a young one and (not yet having undergone its baptism of
fire) only recently regularly fed, it is almost tame and may
be approached without difficulty.   This is why the keeper
just looks round the allotment gardens now and then, and
lets his dogs run about ; for their noses are much more
clever at discovering hidden fur or feathers than his eyes.

In winter if the weather be severe, hares and rabbits
are very bold, and will enter gardens though attached to
dwelling-houses.   Sometimes when a vast double-mound

hedge is grubbed, the ditches each side are left, and the interior space is ultimately converted into an osier-bed, osiers being rather profitable at present. But before these are introduced it is necessary that the ground be well dug up, for it is full of roots and the seeds of weeds, which perhaps have lain dormant for years, but now spring up in wonderful profusion. In consideration of cleansing the soil, and working it by digging, burning the weeds and rubbish, etc., the farmer allows one or two of his labourers to use it as a garden for the growth of potatoes, free of rent, for say a couple of years. Potatoes are a crop which flourish in fresh-turned soil, and so they do very well over the arrangement. But unfortunately as they dig and weed, etc., in the evenings after regular work, they have an excuse for their presence in the fields, and perhaps near preserves at a tempting period of the twenty-four hours. The keeper, in short, is quite aware that some sly poaching goes on in this way.

Another cause of unpleasantness between him and the cottagers arises from the dogs they maintain; generally curs, it is true, and not to all appearance capable of harm. But in the early summer a mongrel cur can do as much mischief as a thoroughbred dog. Young rabbits are easily overtaken when not much larger than rats, and at other seasons, when the game has grown better able to take care of itself, any kind of dog rambling loose in the woods and copses frightens it and unsettles it to an annoying degree.

Consequently, when a dog once begins to trespass, it is pretty sure to disappear for good—it is not necessary to indicate how—and though no actual evidence can be got against the keeper, he is accused of the destruction of the ugly, ill-bred 'pet.' If a dog commences to hunt on his own account he can only be broken off the habit by the utmost severity; and so it sometimes happens that other dogs besides those of the cottagers come to an untimely end by shot and gin.

The keeper, being a man with some true sense of sport, dislikes shooting dogs, though compelled to do so occasionally; he never fires at his own, and candidly admits that he hates to see a sportsman give way to anger in that manner. The custom of 'peppering' with shot a dog for disobedience or wildness, which was once very common in the field, has however gone a great deal into disuse.

Shepherds, who often have to visit their flocks in the night—as at this season of the year, while lambing is in progress—who, in fact, sometimes sleep in the fields in a little wooden house on wheels built for the purpose, are strongly suspected of tampering with the hares scampering over the turnips by moonlight. At harvest-time many strange men come into the district for the extra wages of reaping. They rarely take lodgings—which, indeed, they might find some difficulty in obtaining—but in the warm summer weather sleep in the outhouses and sheds, with the permission of the owner. Others camp in the open in the corner of a meadow, where the angle made by meeting hedges protects them from the wind, crouching round the embers of the fire which boils the pot and kettle. This influx of strangers is not without its attendant anxiety to the keeper, who looks round now and then to see what is going on.

Despite the ill-will in their hearts, the labourers are particularly civil to the keeper ; he is, in fact, a considerable employer of labour—not on his own account—but in the woods and preserves. He can often give men a job in the dead of winter, when farm work is scarce and the wages paid for it are less ; such as hedge-cutting, mending the gaps in the fences, cleaning out ditches or the watercourses through the wood.

Then there is an immense amount of ferreting to be done, and there is such an instinctive love of sport in every man's breast, that to assist in this work is almost an ambition ; besides which, no doubt the chances afforded of an occasional private 'bag' form a secret attraction. One would imagine that there could be but little pleasure in crouching all day in a ditch, perhaps ankle-deep in ice-cold water, with flakes of snow driving in the face, and fingers numbed by the biting wind as it rushes through the bare hawthorn bushes, just to watch a rabbit jump out of a hole into a net, and to break his neck afterwards. Yet so it is ; and some men become so enamoured of this slow sport as to do nothing else the winter through ; and as of course their employment depends entirely upon the will of the keeper, they are anxious to conciliate him.

Despite therefore of missing cur dogs and straying cats which never return, the keeper is treated with marked deference by the cottagers. He is, nevertheless, fully aware of the concealed ill-will towards him ; and perhaps this knowledge has contributed to render him more morose, and sharper of temper, and more suspicious of human nature than he would otherwise have been ; for it never improves a man's character to have to be constantly watching his fellows.

The streams are no more sacred from marauders than the woods and preserves. The brooks and upper waters are not so full of fish as formerly, the canal into which they fall being netted so much ; and another cause of the diminution is the prevalence of fish-poaching, especially for jack, during the spawning season and afterwards. Though the keepers can check this within their own boundaries, it is not of much use.

Fish-poaching is simple and yet clever in its way. In the spawning time jack fish, which at other periods are apparently of a solitary disposition, go in pairs, and sometimes in trios, and are more tame than usual. A long slender ash stick is selected, slender enough to lie light in the hand and strong enough to bear a sudden weight. A loop and running noose are formed of a piece of thin copper wire, the other end of which is twisted round and firmly attached to the smaller end of the stick. The loop is adjusted to the size of the fish—it should not be very much larger, else it will not draw up quick enough, nor too small, else it may touch and disturb the jack. It does not take much practice to hit the happy medium. Approaching the bank of the brook quietly, so as not to shake the ground, to the vibrations of which fish are peculiarly sensitive, the poacher tries if possible to avoid letting his shadow fall across the water.

Some persons' eyes seem to have an extraordinary power of seeing through water, and of distinguishing at a glance a fish from a long swaying strip of dead brown flag, or the rotting pieces of wood which lie at the bottom. The ripple of the breeze, the eddy at the curve, or the sparkle of the sunshine cannot deceive them ; while others, and by far the greater number, are dazzled and see nothing. It is astonishing how few persons seem to have the gift of sight when in the field.

The poacher, having marked his prey in the shallow yonder, gently extends his rod slowly across the water three or four yards higher up the stream, and lets the wire noose sink without noise till it almost or quite touches the bottom. It is easier to guide the noose to its destination when it occasionally touches the mud, for refraction distorts the true position of objects in water, and accuracy is important. Gradually the wire swims down with the current, just as if it were any ordinary twig or root carried

along, such as the jack is accustomed to see, and he therefore feels no alarm. By degrees the loop comes closer to
the fish, till with steady hand the poacher slips it over the
head, past the long vicious jaws and gills, past the first
fins, and pauses when it has reached a place corresponding
to about one-third of the length of the fish, reckoning from
the head. That end of the jack is heavier than the other,
and the 'lines' of the body are there nearly straight.
Thus the poacher gets a firm hold—for a fish, of course,
is slippery—and a good balance. If the operation is performed gently the jack will remain quite still, though the
wire rubs against his side : silence and stillness have such
a power over all living creatures. The poacher now clears
his arm and, with a sudden jerk, lifts the fish right out of
the stream and lands him on the sward.

So sharp is the grasp of the wire that it frequently
cuts its way through the scales, leaving a mark plainly
visible when the jack is offered for sale. The suddenness
and violence of the compression seem to disperse the
muscular forces, and the fish appears dead for the moment.
Very often, indeed, it really is killed by the jerk. This
happens when the loop either has not passed far enough
along the body or has slipped and seized the creature just
at the gills. It then garottes the fish. If, on the other
hand, the wire has been passed too far towards the tail, it
slips off that way, the jack falling back into the water
with a broad white band where the wire has scraped the

scales. Fish thus marked may not unfrequently be seen in the stream. The jack, from its shape, is specially liable to capture in this manner ; long and well balanced, the wire has every chance of holding it. This poaching is always going on ; the implement is so easily obtained and concealed. The wire can be carried in the pocket, and the stick may be cut from an adjacent copse.

The poachers observe that after a fish has once escaped from an attempt of the kind it is ever after far more difficult of capture. The first time the jack was still and took no notice of the insidious approach of the wire gliding along towards it ; but the next—unless a long interval elapses before a second trial—the moment it comes near he is away. At each succeeding attempt, whether hurt or not, he grows more and more suspicious, till at last to merely stand still or stop while walking on the bank is sufficient for him ; he is off with a swish of the tail to the deeper water, leaving behind him a cloud, so to say, of mud swept up from the bottom to conceal the direction of his flight. For it would almost seem as if the jack throws up this mud on purpose ; if much disturbed he will quite discolour the brook. The wire does a good deal to depopulate the stream, and is altogether a deadly implement.

But a clever fish-poacher can land a jack even without a wire, and with no better instrument than a willow stick cut from the nearest osier-bed. The willow, or

withy as it is usually called, is remarkably pliant, and can
be twisted into any shape. Selecting a long slender wand,
the poacher strips it of leaves, gives the smaller end a
couple of twists, making a noose and running knot of the
stick itself. The mode of using it is precisely similar to
that followed with a wire, but it requires a little more
dexterity, because, of course, the wood, flexible as it is,
does not draw up so quickly or so closely as the metal,
neither does it take so firm a grip. A fish once caught
by a wire can be slung about almost anyhow, it holds so
tightly. The withy noose must be jerked up the instant
it passes under that part of the jack where the weight of
the fish is balanced—the centre of gravity; if there is an
error in this respect it should be towards the head, rather
than towards the tail. Directly the jack is thrown out
upon the sward he must be seized, or he will slip from the
noose, and possibly find his way back again into the
water. With a wire there is little risk of that; but then
the withy does not cut its way into the fish.

This trick is often accomplished with the common
withy—not that which grows on pollard trees, but in osier-
beds; that on the trees is brittle. But a special kind is
sought for the purpose, and for any other requiring extreme flexibility. It is, I think, locally called the stone
osier, and it does not grow so tall as the common sorts.
It will tie like string. Being so short, for poaching fish it
has to be fastened to a thicker and longer stick, which is

easily done ; and some prefer it to wire because it looks more natural in the water and does not alarm the fish, while, should the keeper be about, it is easily cut up in several pieces and thrown away.   I have heard of rabbits, and even hares, being caught with a noose of this kind of withy, which is as ' tough as wire ;' and yet it seems hardly possible, as it is so much thicker and would be seen. Still, both hares and rabbits, when playing and scampering about at night, are sometimes curiously · heedless, and foolish enough to run their necks into anything.

With such a rude implement as this some fish-poachers will speedily land a good basket of pike.   During the spawning season, as was observed previously, jack go in pairs, and now and then in trios, and of this the poacher avails himself to take more than one at a haul.   The fish lie so close together—side by side just at that time—that it is quite practicable, with care and judgment, to slip a wire over two at once.   When near the bank two may even be captured with a good withy noose : with a wire a clever hand will make a certainty of it.   The keeper says that on one occasion he watched a man operating just without his jurisdiction, who actually succeeded in wiring three jacks at once and safely landed them on the grass. They were small fish, about a pound to a pound and a half each, and the man was but a few minutes in accomplishing the feat.   It sometimes happens that after a heavy flood, when the brook has been thick with suspended mud for several days, so soon as it has gone down fish are more than usually plentiful, as if the flood had brought them up-stream : poachers are then particularly busy.

Fresh fish—that is, those who are new to that particular part of the brook—are, the poachers say, much more easily captured than those who have made it their home for some time.   They are, in fact, more easily discovered ; they have not yet found out all the nooks and corners, the projecting roots and the hollows under the banks, the dark places where a black shadow falls from overhanging trees and is with difficulty pierced even by a practised eye.   They expose themselves in open places, and meet an untimely fate.

Besides pike, tench are occasionally wired, and now and then even a large roach ; the tench, though a bottom fish, in the shallow brooks may be sometimes detected by the eye, and is not a difficult fish to capture.   Every one has heard of tickling trout : the tench is almost equally amenable to titillation.   Lying at full length on the sward, with his hat off lest it should fall into the water, the poacher peers down into the hole where he has reason to think tench may be found.   This fish is so dark in colour when viewed from above that for a minute or two, till the sight adapts itself to the dull light of the water, the poacher cannot distinguish what he is searching for.   Presently, having made out the position of the tench, he slips his bared arm in slowly, and without splash, and finds little or no trouble as a rule in getting his hand close to the fish without alarming it: tench, indeed, seem rather sluggish. He then passes his fingers under the belly and gently rubs it. Now it would appear that he has the fish in his power, and has only to grasp it. But grasping is not so easy ; or

TICKLING TROUT.

rather it is not so easy to pull a fish up through two feet of superincumbent water which opposes the quick passage of the arm. The gentle rubbing in the first place seems to soothe the fish, so that it becomes perfectly quiescent, except that it slowly rises up in the water, and thus enables the hand to get into proper position for the final seizure.

When it has risen up towards the surface sufficiently far—
the tench must not be driven too near the surface, for it
does not like light and will glide away—the poacher
suddenly snaps as it were ; his thumb and fingers, if he
possibly can manage it, closing on the gills. The body
is so slimy and slippery that there alone a firm hold can
be got, though the poacher will often flick the fish out of
water in an instant so soon as it is near the surface.
Poachers evidently feel as much pleasure in practising
these tricks as the most enthusiastic angler using the implements of legitimate sport.

No advantage is thought too unfair to be taken of fish ;
nothing too brutally unsportsmanlike.   I have seen a pike
killed with a prong as he lay basking in the sun at the top
of the water.   A labourer stealthily approached, and
suddenly speared him with one of the sharp points of the
prong or hayfork he carried : the pike was a good-sized
one too.

The stream, where not strictly preserved, is frequently
netted without the slightest regard to season.   The net is
stretched from bank to bank, and watched by one man,
while the other walks up the brook thirty or forty yards,
and drives the fish down the current into the bag.   With
a long pole he thrashes the water, making a good deal of
splash, and rousing up the mud, which fish dislike and
avoid.   The pole is thrust into every hiding-place, and
pokes everything out.   The watcher by the net knows by

the bobbing under of the corks when a shoal of roach and perch, or a heavy pike, has darted into it, and instantly draws the string and makes his haul. In this way, by sections at a time, the brook, perhaps for half a mile, is quite cleared out. Jack, however, sometimes escape ; they seem remarkably shrewd and quick to learn. If the string is not immediately drawn when they touch the net, they are out of it without a moment's delay : they will double back up stream through all the splashing and mud, and some will even slide as it were between the net and the bank if it does not quite touch in any place, and so get away.

In its downward course the brook irrigates many water meadows, and to drive the stream out upon them there are great wooden hatches. Sometimes a gang of men, discovering that there is a quantity of fish thereabouts, will force down a hatch, which at once shuts off or greatly diminishes the volume of water flowing down the brook, and then rapidly construct a dam across the current below it with the mud of the shore. Above this dam they thrash the water with poles and drive all the fish towards it, and then make a second dam above the first so as to enclose them in a short space. In the making of these dams speed is an object, or the water will accumulate and flow over the hatch ; so hurdles are used, as they afford a support to the mud hastily thrown up. Then with buckets, bowls, and 'scoops,' they bale out the water between the two dams, and quickly reduce their prey to wriggling help-

O

lessness. In this way whole baskets full of fish have been taken, together with eels; and nothing so enclosed can escape.

The mere or lake by the wood is protected by sharp stakes set at the bottom, which would tear poachers' nets; and the keeper does not think any attempt to sweep it has been made of late years, it is too well watched.

SETTING A NIGHT LINE.

But he believes that night lines are frequently laid: a footpath runs along one shore for some distance, and gives easy access, and such lines may be overlooked. He is certain that eels are taken in that way despite his vigilance.

Trespassing for crayfish, too, causes much annoyance. I have known men to get bodily up to the waist into the

great ponds, a few of which yet remain, after carp. These
fish have a curious habit of huddling up in hollows under
the banks ; and those who know where these hollows and
holes are situate can take them by hand if they can come
suddenly upon them. It is said that now and then fish
arc raked out of the ponds with a common rake (such as
is used in haymaking) when lying on the mud in winter.

CHAPTER IX.

Guerilla Warfare—Gun Accidents—Black Sheep.

SCARCELY a keeper can be found who has not got one or more tales to tell of encounters with poachers, sometimes of a desperate character. There is a general similarity in most of the accounts, which exhibit a mixture of ferocity and cowardice on the side of the intruders. The following case, which occurred some years since, brings these contradictory features into relief. The narrator was not the owner of the man-trap described previously.

There had been a great deal of poaching before the affray took place, and finally it grew to horse-stealing : one night two valuable horses were taken from the home park. This naturally roused the indignation of the owner of the estate, who resolved to put a stop to it. Orders were given that if shots were heard in the woods the news should be at once transmitted to head-quarters, no matter at what hour of the night.

One brilliant moonlight night, frosty and clear, the gang came again. A messenger went to the house, and, as previously arranged, two separate parties set out to intercept the rascals. The head keeper had one detachment, whose object it was to secure the main outlet from the wood towards the adjacent town—to cut off retreat. The young squire had charge of the other, which, with the under keeper as guide, was to work its way through the

GOING FOR THE POACHERS.

wood and drive the gang into the ambuscade. In the last party were six men and a mastiff dog; four of the men had guns, the gentleman only a stout cudgel.

They came upon the gang—or rather a part of it, for the poachers were somewhat scattered—in a 'drive' which

ran between tall firs, and was deep in shadow. With a
shout the four or five men in the 'drive,' or green lane,
slipped back behind the trees, and two fired, killing the
mastiff dog on the spot and 'stinging' one man in the
legs.   Quick as they were, the under keeper, to use his
own words, 'got a squint of one fellow as I knowed ; and
I lets drive both barrels in among the firs.   But, bless
you ! it were all over in such a minute that I can't hardly
tell 'ee how it were.   Our squire ran straight at 'em ; but
our men hung back, though they had their guns and he
had nothing but a stick.   I just seen him, as the smoke
rose, hitting at a fellow ; and then, before I could step, I
hears a crack, and the squire he was down on the sward.
One of them beggars had come up behind, and swung his
gun round, and fetched him a purler on the back of his
head.   I picked him up, but he was as good as dead, to
look at ;' and in the confusion the poachers escaped.
They had probably been put up to the ambuscade by one
of the underlings, as they did not pass that way, but
seemed to separate and get off by various paths.   The
'young squire' had to be carried home, and was ill for
months, but ultimately recovered.

Not one of the gang was ever captured, notwithstanding that a member of it was recognised.   Next day an
examination of the spot resulted in the discovery of a trail
of blood upon the grass and dead leaves, which proved
that one of them had been wounded at the first discharge.

It was traced for a short distance and then lost. Not till the excitement had subsided did the under keeper find that he had been hit; one pellet had scored his cheek under the eye, and left a groove still visible.

Some time afterwards a gun was picked up in the ferns, all rusty from exposure, which had doubtless been dropped in the flight. The barrel was very short—not more than eighteen inches in length—having been filed off for convenience of taking to pieces, so as to be carried in a pocket made on purpose in the lining of the coat. Now with a barrel so short as that, sport, in the proper sense of the term, would be impossible; the shot would scatter so quickly after leaving the muzzle that the sportsman would never be able to approach near enough. The use of this gun was clearly to shoot pheasants at roost.

The particular keeper in whose shed the man-trap still lies among the lumber thinks that the class of poachers who come in gangs are as desperate now as ever, and as ready with their weapons. Breech-loading guns have rendered such affrays extremely dangerous on account of the rapidity of fire. Increased severity of punishment may deter a man from entering a wood; but once he is there and compromised, the dread of a heavy sentence is likely to make him fight savagely.

The keeper himself is not altogether averse to a little fisticuffing, in a straightforward kind of way, putting powder and shot on one side. He rather relishes what he calls 'leathering' a poacher with a good tough ground-ash stick.   He gets the opportunity now and then, coming unexpectedly on a couple of fellows rabbiting in a ditch, and he recounts the 'leathering' he has frequently ad-

MAN-TRAP.

ministered with great gusto.   He will even honestly admit that on one occasion—just one, not more—he got himself most thoroughly thrashed by a pair of hulking fellows.

'Some keepers,' he says, 'are always summoning

people, but it don't do no good.   What's the use of summoning a chap for sneaking about with a cur dog and a
wire in his pocket?   His mates in the village clubs together and pays his fine, and he laughs at you.   Why,
down in the town there them mechanic chaps have got a
regular society to pay these here fines for trespass, and the
bench they claps it on strong on purpose.   But it ain't no
good ; they forks out the tin, and then goes and haves a
spree at a public.   Besides which, if I can help it, I don't
much care to send a man to gaol—this, of course, is
between you and me—unless he uses his gun.   If he uses
his gun there ain't nothing too bad for him.   But these
here prisons—every man as ever I knowed go to gaol
always went twice, and kept on going.   There ain't
nothing in the world like a good ground-ash stick.   When
you gives a chap a sound dressing with that there article,
he never shows his face in your wood no more.   There's
fields about here where them mechanics goes as regular as
Saturday comes to try their dogs, as they calls it—and a
precious lot of dogs they keeps among 'em.   But they
never does it on this estate : they knows my habits, you
see.   There's less summonses goes up from this property
than any other for miles, and it's all owing to this here
stick.   A bit of ash is the best physic for poaching as I
knows on.'

   I suspect that he is a little mistaken in his belief that
it is the dread of his personal prowess which keeps tres-

passers away—it is rather due to his known vigilance and watchfulness. His rather hasty notions of taking the law into his own hands are hardly in accord with the spirit of the times; but some allowance must be made for the circumstances of his life, and it is my object to picture the man as he is.

There are other dangers from guns beside these. A brown gaiter indistinctly seen moving some distance off in in the tall dry grass or fern—the wearer hidden by the bushes—has not unfrequently been mistaken for game in the haste and excitement of shooting, and received a salute of leaden hail. This is a danger to which sportsman and keeper are both liable, especially when large parties are engaged in rapid firing; sometimes a particular corner gets very 'hot,' being enfiladed for the moment by several guns. Yet, when the great number of men who shoot is considered, the percentage of serious accidents is small indeed; more fatal accidents probably happen through unskilled persons thoughtlessly playing with guns supposed not to be loaded, or pointing them in joke, than ever occur in the field. The ease with which the breech-loader can be unloaded or reloaded again prevents most persons from carrying it indoors charged; and this in itself is a gain on the side of safety, for perhaps half the fatal accidents take place within doors.

In farmsteads where the owner had the right of shooting, the muzzle-loader was—and still is, when not converted

—kept loaded on the rack. The starlings, perhaps, are making havoc of the thatch, tearing out straw by straw, and working the holes in which they form their nests right through, till in the upper story daylight is visible. When the whistling and calling of the birds tell him they are busy above, the owner slips quickly out with his gun, and brings down three or four at once as they perch in a row on the roof-tree. Or a labourer leaves a message that there is a hare up in the meadow or some wild ducks have settled in the brook. But men who have a gun always in their hands rarely meet with a mishap. The starlings, by-the-bye, soon learn the trick, and are cunning enough to notice which door their enemy generally comes out at, where he can get the best shot ; and the moment the handle of that particular door is turned, off they go.

The village blacksmith will tell you of more than one narrow escape he has had with guns, and especially muzzle-loaders, brought to him to repair. Perhaps a charge could not be ignited through the foulness of the nipple, and the breech had to be unscrewed in the vice ; now and then the breech-piece was so tightly jammed that it could not be turned. Once, being positively assured that there was nothing but some dirt in the barrel and no powder, he was induced to place it in the forge fire ; when—bang ! a charge of shot smashed the window, and the burning coals flew about in a fiery shower. In one instance a black-smith essayed to clear out a barrel which had become

choked with a long iron rod made red-hot : the explosion
which followed drove the rod through his hand and into
the wooden wall of the shed.   Smiths seem to have a
particular fondness for meddling with guns, and generally
have one stowed away somewhere.

It was not wonderful that accidents happened with
the muzzle-loaders, considering the manner in which they
were handled by ignorant persons.   The keeper declares
that many of the cottagers, who have an old single-barrel,
ostensibly to frighten the birds from their gardens, do not
think it properly loaded until the ramrod jumps out of
the barrel.   They ram the charge, and especially the
powder, with such force that the rebound sends the rod
right out, and he has seen those who were not cottagers
follow the same practice.   A close-fitting wad, too tight
for the barrel, will sometimes cause the rod to spring high
above the muzzle : as it is pushed quickly down it com-
presses the air in the tube, which expands with a sharp
report and drives the rod out.

Loading with paper, again, has often resulted in
mischief : sometimes a smouldering fragment remains in
the barrel after the discharge, and on pouring in powder
from the flask it catches and runs like a train up to the
flask, which may burst in the hand.   For this reason to
this day some of the old farmers, clinging to ancient
custom, always load with a clay tobacco pipe-bowl, snapped
off from the stem for the purpose.   It is supposed to hold

just the proper charge, and as it is detached from the horn or flask there is no danger of fire being communicated to the magazine ; so that an explosion, if it happened, would do no serious injury, being confined to the loose powder of the charge itself. Paper used as wads will sometimes continue burning for a short time after being blown out of the gun, and may set fire to straw, or even dry grass.

The old folk, therefore, when it was necessary to shoot the starlings on the thatch, or the sparrows and chaffinches which congregate in the rickyards in such extraordinary numbers—in short, to fire off a gun anywhere near inflammable materials—made it a rule to load with green leaves, which would not burn and could do no harm. The ivy leaf was a special favourite for the purpose—the broad-leaved ivy which grows against houses and in gardens—because it is stout, about the right size to double up and fold into a wad, and is available in winter, being an evergreen, when most other leaves are gone. I have seen guns loaded with ivy leaves many times. When a gun gets foul the ramrod is apt to stick tight if paper is used after pushing it home, and unless a vice be handy no power will draw it out. In this dilemma the old plan used to be to fire it into a hayrick, standing at a short distance ; the hay, yielding slightly, prevented the rod from breaking to pieces when it struck.

Most men who have had much to do with guns have

burst one or more. The keeper in the course of years has had several accidents of the kind ; but none since the breech-loader has come into general use, the reason of course being that two charges cannot be inadvertently inserted one above the other, as frequently occurred in the old guns.

I had a muzzle-loading gun burst in my hands some time since : the breech-piece split, and the nipple, hammer, and part of the barrel there blew out. Fortunately no injury was done ; and I should not note it except for the curious effect upon the tympanum of the ear. The first sense was that of a stunning blow on the head ; on recovering from which the distinction between one sound and another seemed quite lost. The ear could not separate or define them, and whether it was a person speaking, a whistle, the slamming of a door, or the neigh of a horse, it was all the same. Tone, pitch, variation there was none. Though perfectly, and in fact painfully, audible, all sounds were converted into a miserable jangling noise, exactly like that made when a wire in a piano has come loose and jingles. This annoying state of things lasted three days, after which it gradually went off, and in a week had entirely disappeared. Probably the sound of the explosion had been much increased by the cheek slightly touching the stock in the moment of firing, the jar of the wood adding to the vibration. This gun belonged to another person, and was caught up, already loaded, to take advantage of a favourable chance ; it is noticeable that half the accidents happen with a strange gun.

Shot plays curious freaks sometimes : I know a case in which a gun was accidentally discharged in a dairy paved as most dairies are with stone flags. The muzzle was pointed downwards at the time—the shot struck the smooth stone floor, glanced off and up, and hit another person standing almost at right angles, causing a painful wound. It is a marvel that more bird-keepers do not get injured by the bursting of the worn-out firelocks used to frighten birds from the seed. Some of these are not only rusty, but so thin at the muzzle as almost to cut the hand if it accidentally comes into contact with any force.

A collection of curious old guns might be made in the villages ; the flint-locks are nearly all gone, but there are plenty of single-barrels in existence and use which were converted from that ancient system. In the farmhouses here and there may be found such a weapon, half a century old or more, with a barrel not quite equal in length to the punt-gun, but so long that, when carried under the arm of a tall man, the muzzle touches the ground where it is irregular in level. It is slung up to the beam across the ceiling with leathern thongs—one loop for the barrel and the other for the stock. It is still serviceable, having been kept dry ; and the owner will tell you that he has brought down pigeons with it at seventy yards.

Every man believes that his particular gun is the best

in the locality to kill. The owner of this cumbrous weapon, if you exhibit an interest in its history, will take you into the fields and point out a spot where forty years ago he or his immediate ancestor shot four or five wild geese at once, resting the barrel on the branch of a tree in the hedge and sending a quarter of a pound of lead whistling among the flock. The spot the wild geese used to visit in the winter is still remembered, though they come there no more ; drains and cultivation having driven them away from that southern district. In the course of the winter, perhaps, a small flock may be seen at a great height passing over, but they do not alight, and in some years are not observed at all.

There is a trick sometimes practised by poachers which enables them to make rabbits bolt from their holes without the assistance of a ferret. It is a chemical substance emitting a peculiar odour, and, if placed in the burrows drives the rabbits out. Chemical science, indeed, has been called to the aid of poaching in more ways than one : fish, for instance, are sometimes poisoned, or killed by an explosion of dynamite. These latter practises have, however, not yet come into general use, being principally employed by those only who have had some experience of mining or quarrying.

There is a saying that an old poacher makes the best gamekeeper, on the principle of setting a thief to catch a thief : a maxim however, of doubtful value, since no other

person could so thoroughly appreciate the tempting opportunities which must arise day after day. That keepers
themselves are sometimes the worst of depredators must
be admitted. Hitherto I have chiefly described the course
of action followed by honest and conscientious men, truly
anxious for their employers' interests, and taking a personal
pride in a successful shooting season. But there exists a
class of keepers of a very different order, who have done
much to bring sport itself into unmerited odium.

The blackleg keeper is often a man of some natural
ability—a plausible, obsequious rascal, quick in detecting
the weak points of his employer's character, and in practising upon and distorting what were originally generous
impulses. His game mainly depends upon gaining the
entire confidence of his master ; and, not being embar-
rassed by considerations of self-esteem, he is not choice
in the use of means to that end. He knows that if he
can thoroughly worm himself into his employer's good
opinion, the unfavourable reports which may be set afloat
against him will be regarded as the mere tittle-tattle of
envy ; for it is often an amiable weakness on the part
of masters who are really attached to their servants to
maintain a kind of partisanship on behalf of those whom
they have once trusted.

Such a servant finds plentiful occasions for dexterously
gratifying the love of admiration innate in us all. The
manliest athlete and frankest amateur—who would blush

P

at the praise of social equals—finds it hard to resist the apparently bluff outspoken applause of his inferiors bestowed on his prowess in field sports, whether rowing or riding, with rod or gun. Of course it frequently happens that the sportsman really does excel as a shot; but that in no degree lessens the insidious effect of the praise which seems extorted in the excitement of the moment, and to come forth with unpremeditated energy.

The next step is to establish a common ground of indignation; for it is to be observed that those who unite in abuse of a third person have a stronger bond of sympathy than those who mutually admire another. If by accident some unfortunate *contretemps* should cause a passing irritation between his master and the owner of a neighbouring estate, the keeper loses no opportunity of heaping coals upon the fire. He brings daily reports of trespass. Now the other party's keepers have been beating a field beyond their boundaries; now they have ferreted a bank to which they have no right. Another time they have prevented straying pheasants from returning to the covers by intercepting their retreat; and a score of similar tricks. Or perhaps it is the master of a pack of hounds against whom insinuations are directed: cubs are not destroyed sufficiently, and the pheasants are eaten daily.

Sometimes it is a tenant-farmer with a long lease, who cannot be quickly ejected, who has to bear the brunt of these attacks. He is accused of trapping hares and rabbits: he sets the traps so close to the preserves that the pheasants are frequently caught and mortally injured ; he is suspected of laying poisoned grain about. Not content with this he carries his malice so far as to cause the grass or other crops in which outlying nests or young broods are sheltered to be cut before it is ripe, with the object of destroying or driving them away ; and he presents the mowers whose scythes mutilate game with a quart of beer as reward, or furnishes his shepherds with lurchers for poaching. He encourages the gipsies to encamp in the neighbourhood and carry on nightly expeditions by allowing them the use of a field in which to put their vans and horses. With such accounts as these, supported by what looks like evidence, the blackleg keeper gradually works his employer into a state of intense irritation, meantime reaping the reward of the incorruptible guardian and shrewd upright servitor.

At the same time, in the haze of suspicion he has created, the rascal finds a cloak for his own misdeeds. These poachers, trespassers, gipsies, foxes, and refractory tenants afford a useful excuse to account for the comparative scarcity of game. 'What on earth has become of the birds, and where the dickens are the hares ?' asks the angry proprietor. In the spring he recollects being shown by the keeper, with modest pride, some hundreds of young pheasants, flourishing exceedingly. Now he finds the broods have strangely dwindled, and he is informed that these enemies against whom all along he has been warned have made short work of them. If this explanation seems scarcely sufficient, there is always some inexplicable disease to bear the blame : the birds had been going on famously when suddenly they were seized by a mysterious epidemic which decimated their numbers.

All this is doubly annoying, because, in addition to the loss of anticipated sport, there has been an exorbitant expenditure. The larger the number of young broods of pheasants early in the year the better for the dishonest keeper, who has more chances of increasing his own profit, both directly and indirectly. In the first place, there is the little business of buying eggs, not without commissions. More profit is found in the supply of food for the birds : extras and petty disbursements afford further room for pickings.

Then, when the game has been spirited away, the keeper's object is to induce his employer to purchase full-grown pheasants—another chance of secret gratuities— and to turn them out for the battue. That institution is much approved of by keepers of this character, for, the pheasants being confined to a small area, there is less personal exertion than is involved in walking over several thousand acres to look after hares and partridges.

By poisoning his master's mind against some one he not only covers these proceedings but secures himself from the explanation which, if listened to, might set matters right. The accused, attempting to explain, finds a strong prejudice against him, and turns away in dudgeon. Such underhand tricks sometimes cause mischief in a whole district. An unscrupulous keeper may set people of all ranks at discord with each other.

In these malpractices, and in the disposal of game which is bulky, he is occasionally assisted by other keepers of congenial character engaged upon adjacent estates. Gentlemen on intimate terms naturally imagine that their keepers mutually assist each other in the detection of poaching—meeting by appointment, for instance, at night, as the police do, to confer upon their beats. When two or three are thus in league it is not difficult for them to dispose of booty ; they quickly get into communication with professional receivers ; and instances have been known in which petticoats have formed a cover for a steady if small illegal transport of dead game over the frontier.

For his own profit a keeper of this kind may indeed be trusted to prevent poaching on the part of other persons, whose gains would be his loss, since there would remain less for him to smuggle. Very probably it may come to be acknowledged on all sides that he is watchful and always about : an admission that naturally tends to raise him in the esteem of his employer.

Those who could tell tales—his subordinate assistants

—are all more or less implicated, as in return for their silence they are permitted to get pickings: a dozen rabbits now and then, good pay for little work, and plenty of beer. If one of them lets out strange facts in his cups, it signifies nothing: no one takes any heed of a labourer's beerhouse talk. The steward or bailiff has strong suspicions, perhaps, but his motions are known, and his prying eyes defeated. As for the tenants, they groan and bear it.

It is to be regretted that now and then the rural policeman becomes an accomplice in these nefarious practices. His position of necessity brings him much into contact with the keepers of the district within his charge. If they are a 'shady' lot, what with plenty of drink, good fellowship, presents of game, and insidious suggestions of profit, it is not surprising that a man whose pay is not the most liberal should gradually fall away from the path of duty. The keeper can place a great temptation in his way—*i.e.* occasional participation in shooting when certain persons are absent: there are few indeed who can resist the opportunity of enjoying sport. The rural constable often has a beat of very wide area, thinly populated: it is difficult to tell where he may be; he has a reasonable pretext for being about at all hours, and it is impossible that he should be under much supervision. Perhaps he may have a taste for dogs, and breed them for sale, if not openly, on the sly. Now the keeper can try these animals or even break them in in a friendly way ; and when once he has committed himself, and winked at what is going on, the constable feels that he may as well join and share altogether. At outlying wayside 'publics' the keeper and the constable may carouse to the top of their bent : the landlord is only too glad to be on good terms with them ; his own little deviations pass unnoticed, and if by accident they are discovered he has a friend at court to give him a good character.

The worthy pair have an engine of oppression in their hands which effectually overawes the cottagers : they can accuse them of poaching ; and if not proceeding to the ultimatum of a summons, which might not suit their convenience, can lay them under suspicion, which may result in notice to quit their cottages, or to give up their allotment gardens ; and a garden is almost as important to a cottager as his weekly wages. In this way a landlord whose real disposition may be most generous may be made to appear a perfect tyrant, and be disliked by the whole locality. It is to the interest of the keeper and the constable to obtain a conviction now and then ; it gives them the character of vigilance.

Sometimes a blackleg keeper, not satisfied with the plunder of the estate under his guardianship must needs encroach on the lands of neighbouring farmers occupying under small owners ; and so further ill-will is caused. In the end an exposure takes place, and the employer finds to his extreme mortification how deeply he has been deceived ; but the discovery may not be made for years. Of course all keepers of this character are not systematically vicious : many are only guilty occasionally, when a peculiarly favourable opportunity offers.

Another class of keeper is rather passively than actively bad. This is the idle man, whose pipe is ever in his mouth and whose hands are always in his pockets. He is often what is called a good-natured fellow—soft-spoken, respectful, and willing ; liked by everybody ; a capital comrade in his own class, and, in fact, with too many friends of a certain set.

Gamekeeping is an occupation peculiarly favourable to loafing if a man is inclined that way. He can sit on the rails and gates, lounge about the preserves, go to sleep on the sward in the shade ; call at the roadside inn, and, leaning his gun against the tree from which the sign hangs, quaff his quart in indolent dignity. By degrees he easily falls into bad habits, takes too much liquor, finds his hands unsteady, becomes too lazy to repress poaching (which is a weed that must be constantly pulled up, or it will grow with amazing rapidity), and finally is corrupted, and shares the proceeds of bolder rascals. His assistants do as they please. He has no control over them : they know too much about him.

It is a curious fact that there are poaching villages and non-poaching villages. Out of a dozen or more parishes

A GOOD-FOR-NOTHING KEEPER.

forming a petty sessional district one or two will become
notorious for this propensity.   The bench never meet
without a case from them, either for actual poaching or
some cognate offence.   The drinking, fighting, dishonesty,
low gambling, seem ceaseless—like breeding like—till the
place becomes a nest of rascality.   Men hang about the
public-houses all day, betting on horses, loitering; a
blight seems to fall upon them, and a bad repute clings
to the spot for years after the evil itself has been
eradicated.

Q

If a weak keeper gets among such a set as this he succumbs; and the same cause hastens the moral decay of the constable. The latter has a most difficult part to maintain. If he is disposed to carry out the strict letter of his instructions, that does not do—there is a prejudice against *too much severity. English feeling is anti-Draconian; and even the respectable inhabitants would rather endure some little rowdyism than witness an over interference with liberty. If the constable is good-natured, and loth to take strong measures, he either becomes a semi-accomplice or sinks to a nonentity. It is difficult to find a man capable of controlling such a class; it requires tact, and something of the gift of governing men.

By contact with bad characters a weak keeper may be contaminated without volition of his own at first: for we know the truthful saying about touching pitch. The misfortune is that the guilty when at last exposed become notorious; and their infamy spreads abroad, smirching the whole class to which they belong. The honest conscientious men remain in obscurity and get no public credit, though they may far outnumber the evil-disposed.

To make a good keeper it requires not only honesty and skill, but a considerable amount of 'backbone' in the character to resist temptation and to control subordinates. The keeper who has gone to the bad becomes one of the most mischievous members of the community : the faithful and upright keeper is not only a valuable servant but a protection to all kinds of property.

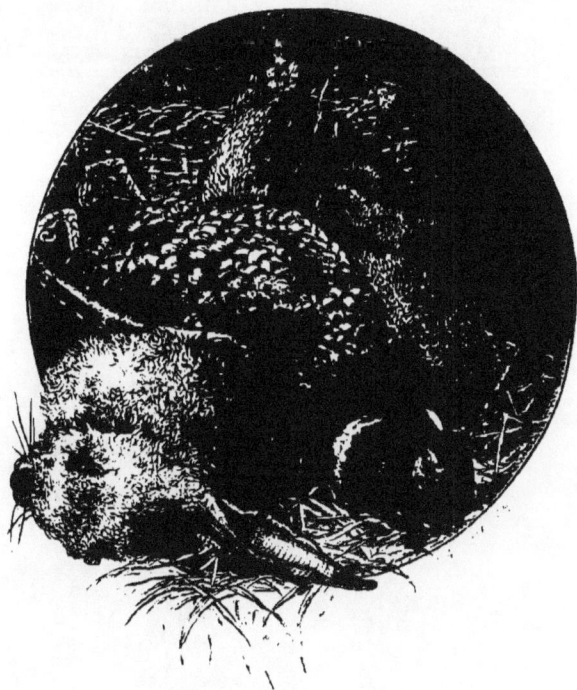

THE END.

Printed by R. & R. CLARK, *Edinburgh.*

www.ingramcontent.com/pod-product-compliance
Lightning Source LLC
Chambersburg PA
CBHW030319270326
41926CB00010B/1429